Applied Data Science Using PySpark

Learn the End-to-End Predictive Model-Building Cycle

Ramcharan Kakarla
Sundar Krishnan
Sridhar Alla

Apress®

Applied Data Science Using PySpark

Ramcharan Kakarla
Philadelphia, PA, USA

Sundar Krishnan
Philadelphia, PA, USA

Sridhar Alla
New Jersey, NJ, USA

ISBN-13 (pbk): 978-1-4842-6499-7
https://doi.org/10.1007/978-1-4842-6500-0

ISBN-13 (electronic): 978-1-4842-6500-0

Managing Director, Apress Media LLC: Welmoed Spahr
Acquisitions Editor: Aditee Mirashi
Development Editor: James Markham
Coordinating Editor: Aditee Mirashi

Cover designed by eStudioCalamar

Cover image designed by Freepik (www.freepik.com)

Distributed to the book trade worldwide by Springer Science+Business Media New York, 1 New York Plaza, Suite 4600, New York, NY 10004-1562, USA. Phone 1-800-SPRINGER, fax (201) 348-4505, e-mail orders-ny@springer-sbm.com, or visit www.springeronline.com. Apress Media, LLC is a California LLC and the sole member (owner) is Springer Science + Business Media Finance Inc (SSBM Finance Inc). SSBM Finance Inc is a **Delaware** corporation.

For information on translations, please e-mail booktranslations@springernature.com; for reprint, paperback, or audio rights, please e-mail bookpermissions@springernature.com.

Apress titles may be purchased in bulk for academic, corporate, or promotional use. eBook versions and licenses are also available for most titles. For more information, reference our Print and eBook Bulk Sales web page at http://www.apress.com/bulk-sales.

Any source code or other supplementary material referenced by the author in this book is available to readers on GitHub via the book's product page, located at www.apress.com/978-1-4842-6499-7. For more detailed information, please visit http://www.apress.com/source-code.

Printed on acid-free paper

Table of Contents

About the Authors

Ramcharan Kakarla is currently Lead Data Scientist at Comcast residing in Philadelphia. He is a passionate data science and artificial intelligence advocate with 6+ years of experience. He graduated as outstanding student with a master's degree from Oklahoma State University with specialization in data mining. Prior to OSU, he received his bachelor's in electrical and Electronics Engineering from Sastra University in India.

He was born and raised in coastal town Kakinada, India. He started his career working as Performance engineer with several fortune 500 clients including StateFarm and British Airways. In his current role he is focused on building data science solutions and frameworks leveraging big data. He has published several award-winning papers and posters in the field of predictive analytics. He served as SAS Global Ambassador for the year 2015. www.linkedin.com/in/ramcharankakarla.

Sundar Krishnan is passionate about Artificial Intelligence and Data Science with more than 5 years of industrial experience. He has tremendous experience in building and deploying customer analytics models and designing machine learning workflow automation. Currently, he is associated with Comcast as a Lead Data Scientist.

Sundar was born and raised in Tamilnadu, India and has a bachelor's degree from Government College of Technology, Coimbatore. He completed his master's at Oklahoma State University, Stillwater. In his spare time, he blogs about his Data science works at Medium website. www.linkedin.com/in/sundarkrishnan1.

Sridhar Alla is founder and CTO of Sas2Py (`www.sas2py.com`) which focuses on automatic conversion of SAS code to Python and on integration with cloud platform services like AWS, Azure, and Google Cloud. His company Bluewhale. one also focuses on using AI to solve key problems, ranging from intelligent email conversation tracking, to solving issues impacting the retail industry, and more. He has deep expertise in building AI-driven big data analytical practices on both public cloud and in-house infrastructures. He is a published author of books and an avid presenter at numerous Strata, Hadoop World, Spark Summit, and other conferences. He also has several patents filed with the US PTO on large-scale computing and distributed systems. He has extensive hands-on experience in most of the prevalent technologies, including Spark, Flink, Hadoop, AWS, Azure, Tensorflow, and others. He lives with his wife Rosie and daughters Evelyn and Madelyn in New Jersey, USA, and in his spare time loves to spend time training, coaching, and attending meetups. He can be reached at sid@bluewhale.one.

About the Technical Reviewer

 **Alessio Tamburro** works currently as Principal Data Scientist in the Enterprise Business Intelligence Innovation Team at Comcast/NBC Universal. Throughout his career in data science research focused teams at different companies, Alessio has gained expertise in identifying, designing, communicating and delivering innovative prototype solutions based on different data sources and meeting the needs of diverse end users spanning from scientists to business stakeholders. His approach is based on thoughtful experimentation and has its roots in his extensive academic research background. Alessio holds a PhD in particle astrophysics from the University of Karlsruhe in Germany and a Master's degree in Physics from the University of Bari in Italy.

Acknowledgments

The completion of this book couldn't have been possible without participation and assistance of many people. I am deeply indebted and appreciate all the support that I received from my friends, family, colleagues during the tough times whose names may not all be enumnerated. Their contributions are sincerely appreciated. I would like to express my deepest appreciation to following individuals who have mentored and supported me:

Dr Goutam Chakraborty (Dr C), Director, MS in Business Analytics OSU for introducing me to the world of data science half decade ago. His passion was contagious, and this book is truly is a product of that passion. I would like to thank my parents, Mr. Mohan Rao Kakarla and Mrs. Venkata Rani for all the enthusiasm and the motivation. I am also grateful to my life partner Mrs. Prasanthi who supported me at all the times. I also would like to thank my friend Mr. Sundar for being a great coauthor to complete this book. I am very thankful to Mr. Sridhar Alla for being a great mentor and supporter.

I also would like to thank Apress team - Aditee Mirashi, Shrikanth Vishwakarma, James Markham for feedback and all the timely efforts in suppoting the completion of the book. A big thanks to Alessio Tamburro, for his technical review and feedback on the book contents.

Heartfelt gratitude to Dr. C, Marina Johnson and Futoshi Yumoto for their continuous support and forewords for this book.

Thank you to all the friends and family who have supported me in this journey.

Ramcharan Kakarla

ACKNOWLEDGMENTS

I wish to express thank you to my wife (Aishwarya) who has supported me throughout the whole process. From reading the drafts of each chapter and providing her views and spending all nighter with me during the book deadlines, I could not have done this without you. Thank you so much, my love.

To my family – Bagavathi Krishnan (Dad), Ganapathy (Mom), Loganathan (Father in law), Manjula (Mother in law), Venkateshwaran (Brother), Manikanda Prabhu (Brother) and Ajey Dhaya Sankar (Brother in law) who are excited about this book as I am and for their motivation and support all day. I would definitely miss the status updates that I provide them on call about this book.

To my data science guru – Dr. Goutam Chakraborty (Dr. C). I still watch his videos to refresh my basics and I built my experience from his teachnigs. For his passion and dedication, I express my gratitude forever.

To my co-author and colleague Ram. We both always had the same passion and we bring the best out of each other. For initiating the idea for the book and working alongside with me to complete the book, I express my thank you.

To Sridhar Alla, for being a mentor in my professional life. For sharing your ideas and supporting our thoughts, I say thank you.

To Alessio Tamburro, for his technical review and feedback on the book contents.

To Dr. C, Marina Johnson and Futoshi Yumoto for their forewords for this book.

To publication coordinators - Aditee Mirashi, Shrikanth Vishwakarma, development editor – James Markham who have assisted us from the start and made this task as easy as possible. To the entire team from the publication who has worked on this book, I say thank you.

Finally, to all the readers who expressed interest to read this book. We hope this book will inspire you to consider your data science passion seriously and build awesome things.

Have fun!

With love,

Sundar Krishnan

I would like to thank my wonderful loving wife, Rosie Sarkaria and my beautiful loving daughters Evelyn & Madelyn for all their love and patience during the many months I spent writing this book. I would also like to thank my parents Ravi and Lakshmi Alla for their blessings and all the support and encouragement they continue to bestow upon me.

Ram and Sundar are simply amazing data scientists who worked hard learning and sharing their knowledge with others. I was lucky to have worked with them and saw them in action. I wish both of them all the very best in all their future endeavors and frankly all credit for this book goes to both of them.

I hope all the readers gain some great insights reading this book and find it one of the books in their personal favorite collection.

Sridhar Alla

Foreword 1

Goutam Chakraborty (Ph.D., University of Iowa) is Professor of Marketing in the Spears School of Business at Oklahoma State University. His research has been published in scholarly journals such as Journal of Interactive Marketing, Journal of Advertising Research, Journal of Advertising, Journal of Business Research, and Industrial Marketing Management, among others. Goutam teaches a variety of courses, including digital business strategy, electronic commerce and interactive marketing, data mining and CRM applications, data base marketing, and advanced marketing research. He has won many teaching awards including "Regents Distinguished Teaching Award" at O.S.U, "Outstanding Direct Marketing Educator Award" given by the Direct Marketing Educational Foundation, "Outstanding Marketing Teacher Award" given by Academy of Marketing Science, and "USDLA Best practice Bronze Award for Excellence in Distance Learning" given by United States Distance Learning Association. He has consulted with numerous companies and has presented programs and workshops worldwide to executives, educators, and research professionals.

No matter where you live in the world today or which industry you work for, your present and future is inundated with messy data all around you. The big questions of the day are "what you are going to do with all these messy data?" and "how would you analyze these messy data using a scalable, parallel processing platform?" I believe this book helps you with answering both of those big questions in an easy to understand style.

I am delighted to write the foreword for the book titled "Applied Data Science Using PySpark" by Ramcharan Kakarla and Sundar Krishnan. Let me say at the outset that I am very positively biased towards the authors because they are graduates of our program at Oklahoma State University and have been through several of my classes. As a professor, nothing pleases me more than when I see my students excel in their lives. The book is a prime example of when outstanding students continue their life long learning and give back to the community of learners their combined experiences in using PySpark for data science.

One of the most difficult things to find in the world of data science books (and there are plenty) is one that takes you through the end-to-end process of conceptualizing, building and implementing data science models in the real world of business. Kakrala and Krishnan have achieved this rarity and done so using PySpark, a widely used platform for data science in the world of big data. So, whether you are a novice planning to enter the world of data science, or you are already in the industry and simply want to upgrade your current skills to using PySpark for data science, this book will be a great place to start.

— Goutam Chakraborty, Ph.D.

Foreword 2

Futoshi Yumoto is actively involved in health outcome research as an affiliated scholar at Collaborative for Research on Outcomes and Metrics (CROM: `https://blogs.commons. georgetown.edu/crom/`), and consistently contributes to scientific advancement through publications and presentations

I really could have used a book like this when I was learning to use PySpark. In my day-to-day work as a data scientist, I plan to use this book as a resource to prototype PySpark codes while I plan to share this book with my team members, to make sure they are easily able to make the most of PySpark from their initial engagement, I can also see this being used as a textbook for a consulting course or certification program.

Currently there are dozens of books with similar titles that are available, but this is the first one I have seen that helps you to set up PySpark environment and execute operation ready codes within a matter of day with sufficient examples, sample data and codes. It leaves the option open for a reader to delve more deeply (and provides up to date references as well as historical ones), while concisely explaining in concrete steps how to apply the program in a variety of problem types/use cases. I have been asked many times for introductory but practical materials to help orient data scientists who needs to transition to PySpark from Python, and this is the first book I am happy to recommend.

This book started as authors project in 2020 and has grown into an organic, but instructive, resource that any data scientist can utilize. Data science is a fast-evolving, multi-dimensional discipline. PySpark has emerged from Spark (define) to help Python users exploit the speed and computational capabilities of Spark. Although this book is focused on Pyspark, through its introduction to both Python and Spark, the authors have crafted a self-directed learning resource that can be generalized to help teach data science principles to individuals at any career stage.

Foreword 3

Marina Evrim Johnson currently works at the Department of Information Management & Business Analytics, Montclair State University. Marina researches the applications of machine learning in the healthcare and operations fields.

You have made a great decision to start learning PySpark with this great book. When Ram and Sundar first mentioned that he was working on a PySpark book, I was thrilled as a university professor teaching Big Data Analytics using PySpark. I had a hard time finding a PySpark book covering Machine Learning and Big Data Analytics from the business perspective. Many Spark books either focus on Scala or Java and also too technical for newbies. Business professionals needed a book that can put PySpark into context using case studies and practical applications. This new book fills this gap and makes it so much easier for people without a technical background to learn the basics of PySpark.

This book is designed for a variety of audiences with different expertise, from newcomers to expert programmers. The book covers many important concepts in the data science and analytics field, including exploratory data analysis, supervised and unsupervised learning methods, and model deployment. Notably, the chapters where authors discuss Machine Leaning flows and automated ML pipelines have many useful tips and tricks that help data science practitioners speed up the model building process.

Given that Python is the most common programming language in the data science and analytics field, PySpark is becoming a default language for big data analytics. With its many packages and seamless connection to Python, PySpark enables data scientists to take advantage of both Python and Spark libraries at the same, providing state-of-the-art analysis and modeling capabilities. I know Ram and Sundar and can guarantee you that they wrote an excellent book. You are going to have a lot of fun reading this book and gain a beneficial skill. I wish you the best of your PySpark learning adventure with this fantastic book.

Introduction

Discover the capabilities of Pyspark and its application in the realm of data science. This comprehensive guide with hand-picked examples of daily use cases will walk you through the end to end predictive model building cycle with the latest techniques and tricks of the trade.

In first 3 chapters, we will get you started with the setting up of the environment, basics of PySpark focusing on data manipulations. We understand feature engineering is where data science professionals spend 70% of their time. As we make you comfortable with the language, we build upon that to introduce you to the mathematical functions available off the shelf. Before we move to the next chapter we will introduce you to the predictive modeling framework. In Chapter 4, we will dive into the art of Variable Selection where we demonstrate various selection techniques available in PySpark. In Chapter 5, 6 & 7, we take you on the journey of machine learning algorithms, implementations and fine-tuning techniques. In addition, we will also talk about different validation metrics and how to use them for picking the best models. Chapter 8 and 9 will walk you through machine learning pipelines, various methods available to operationalize the model and serve it through docker/API. Chapter 10 includes some tricks that can help you optimize your programs and machine learning pipelines.

You will learn the flexibility, advantages and become comfortable using PySpark in data science applications. This book is recommended to data science enthusiasts who want to unleash the power of parallel computing by simultaneously working with big datasets. Highly recommended for professionals who want to switch from traditional languages to open source in big data setting. It's a value add for students who want to work on big data.

What you will learn:

- Overview of end to end predictive model building
- Multiple variable selection techniques & implementations
- Multiple algorithms & implementations
- Operationalizing models
- Data science experimentations & tips

Why this book:

- Comprehensive guide to data science solutions in PySpark

- Desgined Exercises to help you practice your knowledge

- Industry standard methods & procedures all implemented with examples

- Easy to follow structure that can help broad range of audiences (professionals to students)

- Helps you in transitioning your data science solutions from traditional languages to PySpark

- Handpicked Tips & tricks that can help in day to day work

Original picture of authors:

CHAPTER 1

Setting Up the PySpark Environment

The goal of this chapter is to quickly get you set up with the PySpark environment. There are multiple options discussed, so it is up to the reader to pick their favorite. Folks who already have the environment ready can skip to the "Basic Operations" section later in this chapter.

In this chapter, we will cover the following topics:

- Local installation using Anaconda

- Docker-based installation

- Databricks community edition

Local Installation using Anaconda

Step 1: Install Anaconda

The first step is to download Anaconda here: `https://www.anaconda.com/pricing`. Individuals can download the individual edition (free of cost). We recommend using the latest version of Python (Python 3.7) since we use this version in all our examples. If you have an alternate version installed, please make sure to change your code accordingly. Once you install Anaconda, you can run the following command in the Terminal/ Anaconda Prompt to ensure the installation is complete.

© Ramcharan Kakarla, Sundar Krishnan and Sridhar Alla 2021
R. Kakarla et al., *Applied Data Science Using PySpark*, https://doi.org/10.1007/978-1-4842-6500-0_1

Note   Windows users should use Anaconda Prompt not Command Prompt. This option will be available after you install Anaconda.

```
conda info
```

The output should look like the following.

Output shown in Terminal/ Anaconda Prompt

```
(base) ramcharankakarla@Ramcharans-MacBook-Pro % conda info

     active environment : base
    active env location : /Applications/anaconda3
            shell level : 1
       user config file : /Users/ramcharankakarla/.condarc
 populated config files : /Users/ramcharankakarla/.condarc
          conda version : 4.8.3
    conda-build version : 3.18.11
         python version : 3.8.3.final.0
       virtual packages : __osx=10.15.5
       base environment : /Applications/anaconda3   (read only)
           channel URLs : https://repo.anaconda.com/pkgs/main/osx-64
                          https://repo.anaconda.com/pkgs/main/noarch
                          https://repo.anaconda.com/pkgs/r/osx-64
                          https://repo.anaconda.com/pkgs/r/noarch
          package cache : /Applications/anaconda3/pkgs
                          /Users/ramcharankakarla/.conda/pkgs
       envs directories : /Users/ramcharankakarla/.conda/envs
                          /Applications/anaconda3/envs
               platform : osx-64
             user-agent : conda/4.8.3 requests/2.24.0 CPython/3.8.3 Darwin/
                          19.5.0 OSX/10.15.5
                UID:GID : 501:20
             netrc file : None
           offline mode : False
```

Note   Every conda operation starts with the keyword conda.

For more information on conda operations, you can web search for "CONDA CHEAT SHEET" and follow the documentation provided here: `https://docs.conda.io/`.

Step 2: Conda Environment Creation

In general, programming languages have different versions. When we create a program on a specific version and try running the same program on another version, the chances of the program running successfully can vary, depending on the changes made between versions. This can be annoying. Programs developed on an older version may not run well on the newer version because of dependencies. This is where a virtual environment becomes useful. These environments keep these dependencies in separate sandboxes. This makes it easier to run programs or applications in the corresponding environments. The Anaconda tool is an open source platform where you can manage and run your Python programs and environments. You can have multiple environments with different versions being used. You can switch between environments. It is good practice to check which environment you are currently in before executing any program or application.

We would like to focus on a specific command in the conda cheat sheet for environment creation. The syntax is provided here:

```
conda create --name ENVNAME python=3.7
```

Before we create the environment, we are going to take a look at another conda command. We promise it will make sense later why we need to look at this command now.

```
conda env list
```

Once you run this command, you get the following output.

Environment list before creation

```
# conda environments:
#
base                     *  /Applications/anaconda3
```

The *base* environment exists by default in Anaconda. Currently, the *base* is the active environment, which is represented by the small asterisk (*) symbol to its right. What does this mean? It means that whatever conda operations we perform now will be carried out in the default base environment. Any package install, update, or removal will happen in *base*.

Now that we have a clear understanding about the *base* environment, let us go ahead and create our own PySpark environment. Here, we will replace the ENVNAME with a name (pyspark_env) that we would like to call the environment.

```
conda create --name pyspark_env python=3.7
```

When it prompts for a user input, type y and press *Enter*. It is good practice to type these codes rather than copying them from the text, since the hyphen would create an issue, saying either *"CondaValueError: The target prefix is the base prefix. Aborting"* or *"conda: error: unrecognized arguments: --name".* After the command is successfully executed, we will perform the conda environment list again. This time, we have a slightly different output.

Conda environment right after creation

```
# conda environments:
#
base                     *  /Applications/anaconda3
pyspark_env                 /Users/ramcharankakarla/.conda/envs/pyspark_env
```

We notice a new environment *pyspark_env* is created. Still, *base* is the active environment. Let us activate our new environment *pyspark_env* using the following command.

Conda environment after change

```
conda activate pyspark_env
 # conda environments:
#
base                        /Applications/anaconda3
pyspark_env              *  /Users/ramcharankakarla/.conda/envs/pyspark_env
```

Going forward, all the `conda` operations will be performed in the new PySpark environment. Next, we will proceed with Spark installation. Observe that "*" indicates the current environment.

Step 3: Download and Unpack Apache Spark

You can download the latest version of Apache Spark at `https://spark.apache.org/downloads.html`. For this entire book, we will use Spark 3.0.1. Once you visit the site, you can pick a Spark release and download the file (Figure 1-1).

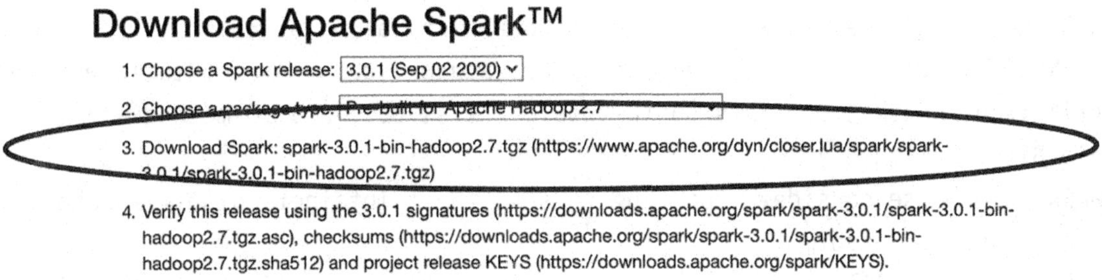

Download Apache Spark™

1. Choose a Spark release: 3.0.1 (Sep 02 2020) ⌄
2. Choose a package type: Pre-built for Apache Hadoop 2.7
3. Download Spark: spark-3.0.1-bin-hadoop2.7.tgz (https://www.apache.org/dyn/closer.lua/spark/spark-3.0.1/spark-3.0.1-bin-hadoop2.7.tgz)
4. Verify this release using the 3.0.1 signatures (https://downloads.apache.org/spark/spark-3.0.1/spark-3.0.1-bin-hadoop2.7.tgz.asc), checksums (https://downloads.apache.org/spark/spark-3.0.1/spark-3.0.1-bin-hadoop2.7.tgz.sha512) and project release KEYS (https://downloads.apache.org/spark/KEYS).

Figure 1-1. *Spark download*

Unzip it, and optionally you can move it to a specified folder.
Mac/Linux Users

```
tar -xzf spark-3.0.1-bin-hadoop2.7.tgz
mv spark-3.0.1-bin-hadoop2.7 /opt/spark-3.0.1
```

Windows Users

```
tar -xzf spark-3.0.1-bin-hadoop2.7.tgz
move spark-3.0.1-bin-hadoop2.7 C:\Users\username\spark-3.0.1
```

Step 4: Install Java 8 or Later

You can check for an existing version of Java in your system by typing the following in Terminal/Anaconda Prompt.

Java version check

```
java -version
 java version "14.0.2" 2020-07-14
Java(TM) SE Runtime Environment (build 14.0.2+12-46)
Java HotSpot(TM) 64-Bit Server VM (build 14.0.2+12-46, mixed mode, sharing)
```

You need to make sure that the Java version is 8 or later. You can always download the latest version at https://www.oracle.com/java/technologies/javase-downloads.html. It is recommended to use Java 8 JDK. When you use a Java version later than 8 you might get the following warnings when you launch PySpark. These warnings can be ignored.

```
WARNING: An illegal reflective access operation has occurred
WARNING: Illegal reflective access by org.apache.spark.unsafe.Platform
(file:/opt/spark-3.0.1/jars/spark-unsafe_2.12-3.0.1.jar) to method java.
nio.Bits.unaligned()
WARNING: Please consider reporting this to the maintainers of org.apache.
spark.unsafe.Platform
WARNING: Use --illegal-access=warn to enable warnings of further illegal
reflective access operations
WARNING: All illegal access operations will be denied in a future release
```

Mac/Linux users can now proceed to Step 5, and Windows users can jump to Step 6 from here.

Step 5: Mac & Linux Users

In Step 3, we completed the Spark download. Now, we must update the *~/.bash_profile* or *~/.bashrc* files to find the Spark installation. Let us use the *bash_profile* file in our example here. In Terminal, type the following code to open the file:

```
vi ~/.bash_profile
```

Once you are inside the file, type i to insert/configure the environment variables by adding the following lines:

```
export SPARK_HOME=/opt/spark-3.0.1
export PATH=$SPARK_HOME/bin:$PATH
export PYSPARK_PYTHON=python3
```

After adding these configurations, press *ESC* followed by :wq! and press *Enter* to save the file. Run the following command in Terminal to apply these changes in our *pyspark_env* environment:

```
source ~/.bash_profile
```

If you restart your Terminal, make sure to change the conda environment to *pyspark_env* to run PySpark. You can now skip Step 6 and jump to Step 7.

Step 6: Windows Users

In Step 3, we completed the Spark download. Now, we need to do a couple of things to make sure our PySpark installation works as intended. Currently, if you made the move option in Step 3, your Spark folder should exist in the path provided here:

```
C:\Users\username\spark-3.0.1
```

Navigate to this path via Anaconda Prompt and run the following code:

```
cd C:\Users\username\spark-3.0.1
mkdir hadoop\bin
```

Once the directory is created, we need to download the *winutils.exe* file and add it to our Spark installation. Why *winutils.exe*? Without the file, when you invoke PySpark, you might get the following error:

```
"ERROR Shell: Failed to locate the winutils binary in the hadoop binary
path....."
```

In addition, there could be errors in the future when you try to use the spark-submit utility. In order to overcome these errors, we will download the *winutils.exe* file: https://github.com/steveloughran/winutils. Based on the hadoop version you installed in Step 3, you might have to pick the appropriate path in this link. Navigate to the bin folder inside the path and click on *winutils.exe* to download the file (Figure 1-2).

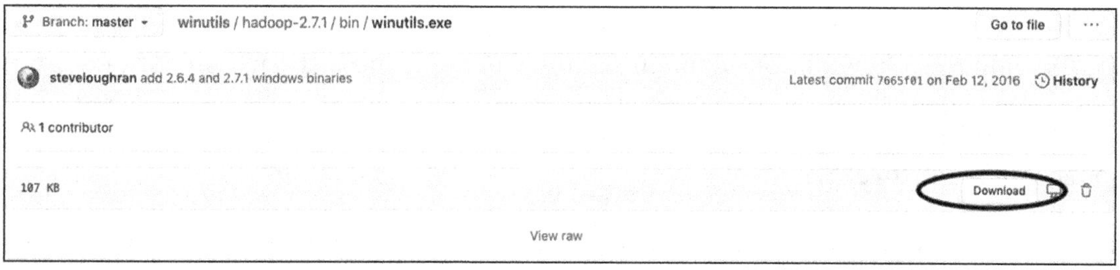

Figure 1-2. *Winutils.exe*

Folks who are using the 2.7 hadoop version can download the *winutils.exe* file directly from here: `https://github.com/steveloughran/winutils/blob/master/hadoop-2.7.1/bin/winutils.exe`. Place the downloaded file in the `hadoop\bin` path, which we created earlier. Alternatively, you can use the following command to perform the same operation:

```
move C:\Users\username\Downloads\winutils.exe C:\Users\username\
spark-3.0.1\hadoop\bin
```

That's it. We are almost done. The final step is to configure the environment variables. You can search for *"Environment Variables"* in Windows search or you can find them in *System Properties* ➤ *Advanced* ➤ *Environment Variables*. Here, we will add a few variables in the *User variables* tab. Click on *New* and you should see the option shown in Figure 1-3.

Figure 1-3. *Add new user variables*

We need to add three things: SPARK_HOME, HADOOP_HOME, *and* PATH.
SPARK_HOME

```
Variable name - SPARK_HOME
Variable value - C:\Users\username\spark-3.0.1
```

HADOOP_HOME

Variable name – HADOOP_HOME
Variable value – C:\Users\username\spark-3.0.1\hadoop

If there is an existing PATH variable in the *User variables* tab, then click on Edit and add the following:

C:\Users\username\spark-3.0.1\bin

Well, that is all. You can now run PySpark. To apply all these environment changes, you need to restart the Anaconda Prompt and change the conda environment to *pyspark_env*.

Step 7: Run PySpark

You are all set now. You need to make sure that the conda environment is *pyspark_env* before calling pyspark. Just invoke pyspark from the Terminal/Anaconda Prompt. You should see the following pop up.

Run Pyspark in Terminal/Anaconda Prompt

```
Python 3.7.9 (default, Aug 31 2020, 07:22:35)
[Clang 10.0.0 ] :: Anaconda, Inc. on darwin
Type "help", "copyright", "credits" or "license" for more information.
WARNING: An illegal reflective access operation has occurred
WARNING: Illegal reflective access by org.apache.spark.unsafe.Platform
(file:/opt/spark-3.0.1/jars/spark-unsafe_2.12-3.0.1.jar) to constructor
java.nio.DirectByteBuffer(long,int)
WARNING: Please consider reporting this to the maintainers of org.apache.
spark.unsafe.Platform
WARNING: Use --illegal-access=warn to enable warnings of further illegal
reflective access operations
WARNING: All illegal access operations will be denied in a future release
20/09/11 22:02:30 WARN NativeCodeLoader: Unable to load native-hadoop
library for your platform... using builtin-java classes where applicable
Using Spark's default log4j profile: org/apache/spark/log4j-defaults.
properties
```

```
Setting default log level to "WARN".
To adjust logging level use sc.setLogLevel(newLevel). For SparkR, use
setLogLevel(newLevel).
Welcome to

      ____              __
     / __/__  ___ _____/ /__
    _\ \/ _ \/ _ `/ __/  '_/
   /__ / .__/\_,_/_/ /_/\_\   version 3.0.1
      /_/

Using Python version 3.7.9 (default, Aug 31 2020 07:22:35)
SparkSession available as 'spark'.
>>>
```

Cool! We would like to extend the setup a little further to work with Jupyter IDE.

Step 8: Jupyter Notebook Extension

For this, you might have to install Jupyter in your *pyspark_env* environment. Let us use the following code to perform the operation.

Mac/Linux users

```
conda install jupyter
```

You need to add the following two lines of code in your *~/.bash_profile* or *~/.bashrc* setup:

```
export PYSPARK_DRIVER_PYTHON=jupyter
export PYSPARK_DRIVER_PYTHON_OPTS='notebook'
```

That's it. Make sure to apply the current changes in the environment by using the appropriate code:

```
source ~/.bash_profile
source ~/.bashrc
```

Going forward, whenever you invoke pyspark, it should automatically invoke a Jupyter Notebook session. If the browser did not open automatically, you can copy the link and paste it in the browser.

Jupyter invocation

```
[I 22:11:33.938 NotebookApp] Writing notebook server cookie secret to /
Users/ramcharankakarla/Library/Jupyter/runtime/notebook_cookie_secret
[I 22:11:34.309 NotebookApp] Serving notebooks from local directory: /
Applications
[I 22:11:34.309 NotebookApp] Jupyter Notebook 6.1.1 is running at:
[I 22:11:34.309 NotebookApp] http://localhost:8888/?token=2e2d4e2c7a881de72
23ea48aed4e696264ef9084c2d36ccf
[I 22:11:34.310 NotebookApp]  or http://127.0.0.1:8888/?token=2e2d4e2c7a881
de7223ea48aed4e696264ef9084c2d36ccf
[I 22:11:34.310 NotebookApp] Use Control-C to stop this server and shut
down all kernels (twice to skip confirmation).
[C 22:11:34.320 NotebookApp]
```

To access the notebook, open this file in a browser:

```
file:///Users/ramcharankakarla/Library/Jupyter/runtime/nbserver-45444-
open.html
```

Or copy and paste one of these URLs:

```
    http://localhost:8888/?token=2e2d4e2c7a881de7223ea48aed4e696264ef90
    84c2d36ccf
```

Windows users

You need to install a package using the Anaconda Prompt:

```
python -m pip install findspark
```

That's it. Going forward, whenever you invoke `Jupyter Notebook` from Anaconda Prompt, it should automatically invoke a Jupyter Notebook session. If the browser did not open automatically, you can copy the link and paste it in the browser, as shown in Figure 1-10. In the browser, you need to select a *New Python 3* notebook. To test the PySpark installation in Windows, run the following code in Jupyter:

```
## Initial check
import findspark
findspark.init()
```

```
## Run these codes when the initial check is successful
import pyspark
from pyspark.sql import SparkSession
spark = SparkSession.builder.getOrCreate()
df = spark.sql("select 'spark' as hello ")
df.show()
```

Well, that is all with the setup. You can now skip to the "Basic Operations" section later in this chapter.

Docker-based Installation

Believe it or not, but a Docker-based installation is very simple when compared to a local installation. Some readers might just be getting started with Docker, whereas a few might be proficient. In either case, it does not matter. This section will introduce Docker to readers and show you how to use PySpark with Docker. One thing we would like to make clear before we go further: Docker by itself is a vast resource, and this book cannot cover it all. You can read online articles or buy a separate book that teaches you about Docker. Here, we will cover the basic things that you need to know to use Docker for our purposes. Rest assured, once you get the hang of Docker, you will prefer to use it over anything else. Proficient users can skip directly to the Docker image installation discussed later in this section.

Why Do We Need to Use Docker?

Think of a scenario where you just created an application using Python 2.7. All of a sudden, the admin in your college/company says that they are going to switch to Python 3.5 or later. Now, you have to go all the way back to test your code's compatibility with the new Python version, or you might have to create a separate environment to run your application standalone. Well, that is going to be frustrating and a hassle given the amount of time you need to spend to make your application run as usual. Docker comes in handy in such a situation. Once you create a Docker image, changes happening in the outside environment do not matter. Your application will work in any place Docker exists. This makes Docker extremely powerful, and it becomes a crucial application for data science. As a data scientist, you might run into scenarios where you need to develop and deploy models in production. If a small change in the production environment occurs, it could possibly break your code and your application. Thus, using Docker will make your life better.

What Is Docker?

Docker is a platform as a service (PaaS) product. You create your own platform on which to run your application. Once the Docker platform is created, it can be hosted any place where you have Docker. Docker sits on top of a host OS (Mac/Windows/Linux, etc.). After installing the Docker application, you can create images, which are called Docker images. From these images, you can create a container, which is an isolated system. Inside a container, you host your application and install all the dependencies (*bins/libs*) for your application. This is shown in Figure 1-4.

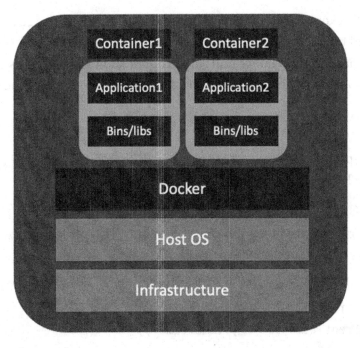

Figure 1-4. *Docker*

Wait a second! It sounds similar to a virtual machine. Well, it is not. Docker differs in its internal working mechanism. Each virtual machine needs a separate OS, whereas in Docker the OS can be shared. In addition, virtual machines use more disk space and have higher overhead utilization when compared to Docker. Let us learn how to create a simple Docker image.

Create a Simple Docker Image

Before creating an image, open a Docker account, as follows:

- Create a *Docker Hub* account at `https://hub.docker.com/signup`.

- Download *Docker Desktop* from `https://www.docker.com/get-started`.

- Install *Docker Desktop* and sign in using the *Docker Hub* user ID/password.

To create a Docker image, you need a `Dockerfile.` The `Dockerfile` commands are available here: `https://docs.docker.com/engine/reference/builder/`. To try it out, we will create a basic Python image and run an application inside it. Let's get started. Using a text editor, open a new file named *Dockerfile* and copy/paste the following commands:

```
FROM python:3.7-slim
COPY test_script.py /
RUN pip install jupyter
CMD ["python", "./test_script.py"]
```

The *test_script.py* file has the following commands. This file should exist in the same location as *Dockerfile* for this code to work:

```
import time
print("Hello world!")
print("The current time is - ", time.time())
```

What is happening here? We are using a base image, which is Python 3.7, slim version. Next, we copy the local file *test_script.py* to Docker. In addition, we show a quick way to install a package within the Docker instance, and finally we execute the *test_script.py* file. That's all. We can now create our Docker image using Docker commands.

All the Docker commands are provided here: `https://docs.docker.com/engine/reference/commandline/docker/`. For this simple image creation, we will look at two commands, as follows:

- `docker build` – Create image

- `docker run` – Run image

From your Terminal/Command Prompt, navigate to the folder where *Dockerfile* resides and run the following command:

```
docker build -t username/test_image .
docker run username/test_image
```

The first command creates the image. The `-t` is used to assign a tag for the image. You can replace the username with your Docker Hub username. Make sure to include the `period (.)` after the tag name. This will automatically consume `Dockerfile` in the location to create the image. The second command is used to run the image. When executed, you should get an output saying *"Hello world!. The current time is—blah. blah."*

We now have a basic understanding of Docker, so let us go ahead and download PySpark Docker.

Download PySpark Docker

- Go to `https://hub.docker.com/` and search for the *PySpark* image.

- Navigate to the image shown in Figure 1-5.

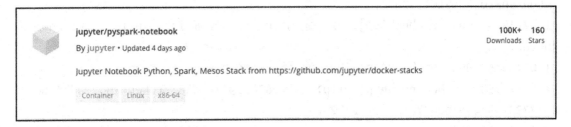

Figure 1-5. PySpark image

- From the Terminal/Command Prompt, type the following code and press *Enter*:

    docker pull jupyter/pyspark-notebook

- Once the image is completely downloaded, you can run the following code to ensure the image exists. One thing to notice here is that there is a *latest* tag associated with this image. If you have multiple images with same name, the tag could be used to differentiate versions of those images.

```
docker images
```

Step-by-Step Approach to Understanding the Docker PySpark run Command

1. Let us start with the basic Docker run command.

Listing 1-1. Docker run base

```
docker run jupyter/pyspark-notebook:latest

(base) ramcharankakarla@Ramcharans-MacBook-Pro ~ % docker run jupyter/
pyspark-notebook:latest
Executing the command: jupyter notebook
[I 02:19:50.508 NotebookApp] Writing notebook server cookie secret to /
home/jovyan/.local/share/jupyter/runtime/notebook_cookie_secret
[I 02:19:53.064 NotebookApp] JupyterLab extension loaded from /opt/conda/
lib/python3.8/site-packages/jupyterlab
[I 02:19:53.064 NotebookApp] JupyterLab application directory is /opt/
conda/share/jupyter/lab
[I 02:19:53.068 NotebookApp] Serving notebooks from local directory: /home/
jovyan
[I 02:19:53.069 NotebookApp] The Jupyter Notebook is running at:
[I 02:19:53.069 NotebookApp] http://b8206282a9f7:8888/?token=1607408740dd46
b6e3233fa9d45d3a0b278fd282e72d8242
[I 02:19:53.069 NotebookApp]  or http://127.0.0.1:8888/?token=1607408740dd4
6b6e3233fa9d45d3a0b278fd282e72d8242
[I 02:19:53.069 NotebookApp] Use Control-C to stop this server and shut
down all kernels (twice to skip confirmation).
[C 02:19:53.074 NotebookApp]
```

To access the notebook, open this file in a browser:

```
file:///home/jovyan/.local/share/jupyter/runtime/
nbserver-7-open.html
```

Or copy and paste one of these URLs:

```
http://b8206282a9f7:8888/?token=1607408740dd46b6e
3233fa9d45d3a0b278fd282e72d8242
or http://127.0.0.1:8888/?token=1607408740dd46b6e3233
fa9d45d3a0b278fd282e72d8242
```

> When you copy and paste the preceding link in the browser, you won't see any Jupyter session. This is because there is no connection established between Docker and the outside world. In the next step, we will establish that connection using a port.

2. A port can be established with the following command:

    ```
    docker run -p 7777:8888 jupyter/pyspark-notebook:latest
    ```

    We chose two different ports here just for explanatory purposes. After you execute this command, you will have a similar output as that shown in Listing 1-1. When you copy and paste as before, you won't see any session. However, instead of using port 8888, if you use the port 7777, it will work. Try with this link http://localhost:7777/ and you should be able to see a Jupyter session. What is happening here?

    Inside Docker, the Jupyter Notebook session is running in port 8888. However, when we established the Docker connection, we explicitly mentioned that Docker should use port 7777 to communicate with the outside world. Port 7777 becomes the doorway, and that is why we see the output. Now that you are clear about the port syntax, let us assign the same port to both sides, as follows. This time, the link shown should work directly.

    ```
    docker run -p 8888:8888 jupyter/pyspark-notebook:latest
    ```

3. What if you want to access external files within Docker? This is again established using a simple command:

```
docker run -p 8888:8888 -v /Users/ramcharankakarla/demo_
data/:/home/jovyan/work/ jupyter/pyspark-notebook:latest
```

Again, the Jupyter Notebook is available in the link shown from running this code. However, there is something going on here. Let us examine this command in depth. Like the port mapping we saw before, there is left- and right-side mapping, here followed by a symbol -v. So, what is this? The -v denotes the volume to mount on while invoking the docker command. The left side is the local directory that you want to mount when you invoke docker. It means that whatever files and folders are available in the local directory /Users/sundar/ will be available in the Docker path /home/work. Here is the thing though: When you open the Jupyter session, you will see the /work directory, which is basically empty, and not /home/work. That is okay, because we are not invoking it in the interactive/bash mode, which we will discuss in the next step.

4. How do you invoke in an interactive/bash mode? Use the following command:

```
docker run -it -p 8888:8888 -v /Users/ramcharankakarla/
demo_data/:/home/jovyan/work/ jupyter/pyspark-
notebook:latest
```

Okay. You can see two changes in the command. First is the -it tag, which denotes an interactive session is to be invoked. The second is the bash command. This would invoke the shell. When you execute this command, you won't see a Jupyter session link this time. Rather, you will be taken into a Linux shell, where you can execute further commands. This is useful in a lot of ways. You can install packages as we did before in the Linux environment. Navigate to the folder we mapped earlier and then invoke the Jupyter Notebook using the following command:

```
cd /home/work
jupyter notebook
```

You should be able to see the files and folders in the Jupyter session.

5. There are additional tags for space optimization. We recommend you use the following final command when you run Docker:

```
docker run -it --rm --shm-size=1g --ulimit memlock=-1 -p
8888:8888 -v /Users/ramcharankakarla/demo_data/:/home/
jovyan/work/ jupyter/pyspark-notebook:latest
```

The `--rm` tag is used to delete the container after the specified task is complete. This is suitable for short-lived containers. You explicitly say to the Docker Daemon that once this container is finished running, clean up the files and delete them to save disk space. The second tag, `--shm-size,` is used to increase/decrease the shared memory size. For memory-intensive containers, you can use this option to specify the shared memory size, which helps to run the containers faster by giving additional access to allocated memory. The final tag is `--ulimit memlock=-1`, which provides the container an unlimited locked-in-memory address space. This will lock the maximum available shared memory during the container invocation. You don't have to use all these options; however, it is good to know them. Now you have a working PySpark session with Docker.

Databricks Community Edition

This is intended for users who are looking to use Databricks. It will come in handy when we discuss FL flow in Chapter 7 as well. If you already have the Databricks environment set up at your workplace, you can reach out to your admin to create an account for you. For folks who intend to try to install this on their own, use this link to sign up: `https://databricks.com/try-databricks`. After you create your account, you will be provided with couple of options to play with as shown in Figure 1-6.

Select a platform

DATABRICKS PLATFORM - FREE TRIAL

For businesses

- Collaborative environment for Data teams to build solutions together
- Unlimited clusters that can scale to any size, processing data in your own account
- Job scheduler to execute jobs for production pipelines
- Fully collaborative notebooks with multi-language support, dashboards, REST APIs
- Native integration with the most popular ML frameworks (scikit-learn, TensorFlow, Keras,...), Apache SparkTM, Delta Lake, and MLflow
- Advanced security, role-based access controls, and audit logs
- Single Sign On support
- Integration with BI tools such as Tableau, Qlik, and Looker
- 14-day full feature trial (excludes cloud charges)

COMMUNITY EDITION

For students and educational institutions

- Single cluster limited to 6GB and no worker nodes
- Basic notebooks without collaboration
- Limited to 3 max users
- Public environment to share your work

GET STARTED

By clicking "Get Started" for the Community Edition, you agree to the Databricks Community Edition Terms of Service.

GET STARTED ON

  OR

Please note that Azure Databricks is provided by Microsoft and is subject to Microsoft's terms.

By clicking on the "AWS" button to get started, you agree to the Databricks Terms of Service.

Figure 1-6. *Databricks options*

For now, you should be good with the community edition. Once you click on *Get Started,* you should be able to navigate based on the instructions in your email.

Create Databricks Account

- Create an account here: `https://databricks.com/try-databricks`

- You might have to authenticate from the email. Once you have the authentication done, then you are good to use Databricks.

Create a New Cluster

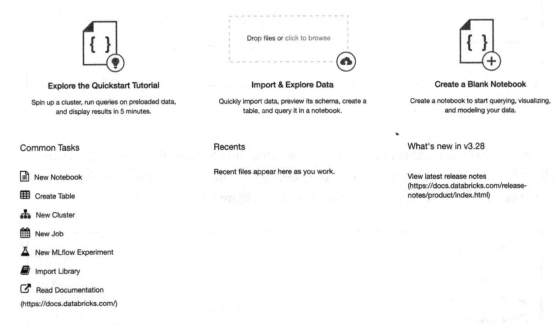

Figure 1-7. *Databricks cluster*

You need to provide a cluster name (*pyspark_env*) and then choose a Databricks runtime version (Figure 1-8). The packages that we use for our data science demonstration are available in the ML runtime version. Let us go ahead and choose the latest 2.4.5 release 6.6 ML version for our task and create the cluster. To read more on the cluster runtime version, you can visit here: `https://docs.databricks.com/release-notes/runtime/releases.html`. Optionally, you can also use the GPU version of the ML, but it is not required.

Figure 1-8. *Cluster creation*

After you click on Create Cluster, it takes some time to build the cluster. Once the cluster is created, you can start using the notebooks. This should be denoted with a green dot before the cluster name you specified (Figure 1-9).

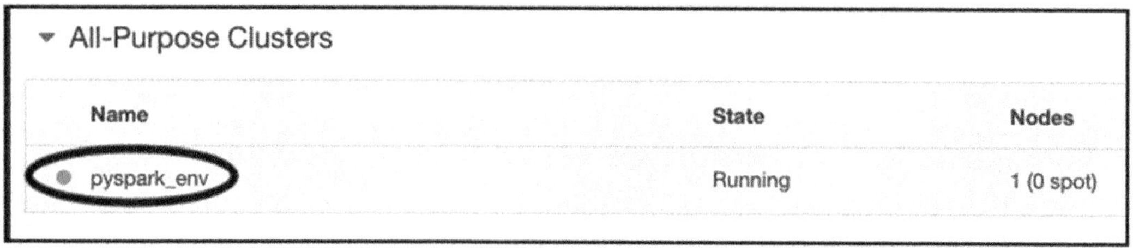

Figure 1-9. *Cluster creation complete*

Create Notebooks

Click on the Databricks icon on the top-left corner to navigate to the Welcome page (Figure 1-7). From there, you can click on *New Notebook* in the Common Tasks section and create a notebook. You need to make sure to specify the cluster name you created in the previous task. This is shown in Figure 1-10.

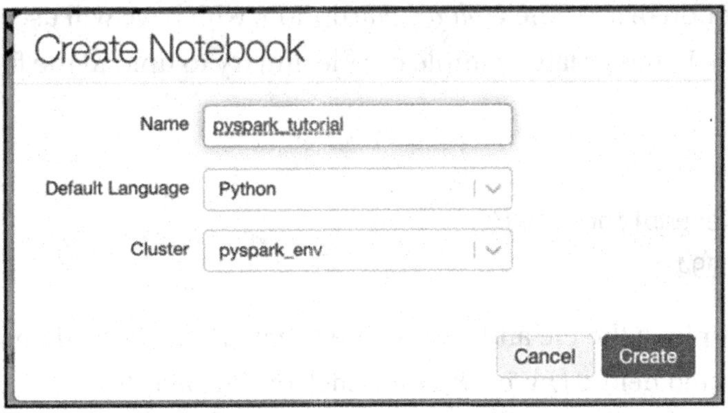

Figure 1-10. *Databricks notebook*

You can type the following code in the notebook for testing purpose and then press *Shift + Enter* or the use the *Run Cell* option in the right corner of the cell.

You are now all set up to run a PySpark notebook in Databricks. Easy, right? But you need to know one more thing before we close this section. How can you import your data into the Databricks environment? We will see that next.

How Do You Import Data Files into the Databricks Environment?

Select the *Data* tab located in the left panel. Click on *Add Data*.

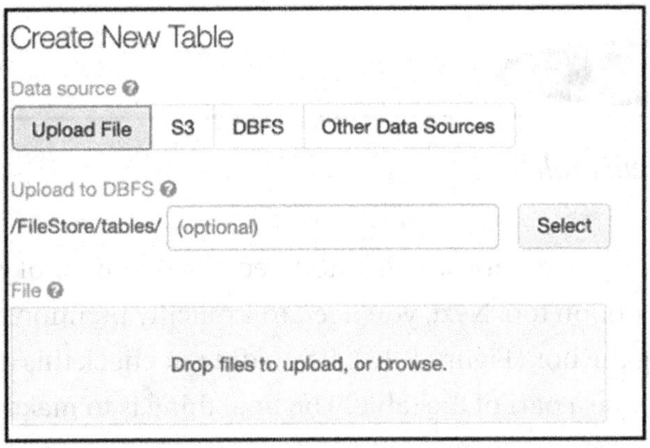

Figure 1-11. *Create new table*

The easiest option of all is the *Upload File* option, which we will use most in this book (Figure 1-11). Let us create a simple csv file and try to upload the file.

```
movie,year
The Godfather,1972
The Shawshank Redemption, 1994
Pulp Fiction, 1994
```

You can now upload the file and choose the *Create Table with UI* option, then select the cluster we created before (*pyspark_env*). Click on *Preview Table.*

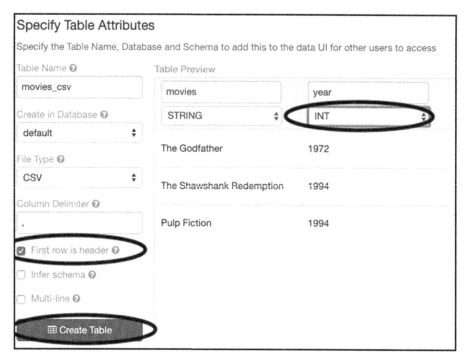

Figure 1-12. *Preview table*

A lot is going on here. Databricks will autodetect the delimiter of your file; however, if it fails you have an option too. Next, you need to explicitly mention that the *First row is header* by checking the box (Figure 1-12). If you did not check this option, then the column names would be a part of the table. The final thing is to make sure the data types are appropriate before you create the table. Column *year* is an integer data type (INT).

Therefore, you need to change it accordingly. By default, Databricks chooses the STRING datatype. Once all the table attributes are specified, you can click on the *Create Table* option. What you will notice here is that it creates a new table in the *default* database.

Let us head back to the notebook using the *Workspace* option on the left panel and try to access this data. Usually, when you access the table using the PySpark SQL option, it creates a PySpark dataframe. We will talk more about this later. For now, let us use the following code to access the data:

```
import pyspark
df = sqlContext.sql("select * from movies_csv")
df.show()
```

You should get the following output (Figure 1-13):

```
+--------------------+----+
|              movies|year|
+--------------------+----+
|       The Godfather|1972|
|The Shawshank Red...|1994|
|        Pulp Fiction|1994|
+--------------------+----+
```

Figure 1-13. *Databricks PySpark output*

Pretty cool, right? You are all set to use the Databricks version.

Basic Operations

Congratulations on setting up your PySpark environment. In this section, we will use some toy codes to play with the environment you just created. All the demonstration shown hereafter should work irrespective of the setup option you chose before.

Upload Data

This section is intended for users who chose either local or Docker-based installation. Databricks users can skip this and proceed to the next operation. Once you invoke the Jupyter Notebook, you have the option to upload data directly to your environment (Figure 1-14). Let us create a simple csv file and name it *movies.csv*.

```
movie,year
The Godfather,1972
The Shawshank Redemption, 1994
Pulp Fiction, 1994
```

Figure 1-14. *Upload files*

Access Data

Once the file is uploaded, jump to your PySpark notebook and run the following code:

```
from pyspark.sql import SparkSession
spark = SparkSession.builder.getOrCreate()
df = spark.read.csv("movies.csv", header=True, sep=',')
df.show()
```

You should see output similar to that in Figure 2-1.

Calculate Pi

This is a common example that you see online or in textbooks. We will calculate the value of pi using 100 million samples to test the PySpark operation. Once the operation is complete, it will provide you the value of Pi. Use the following code:

```
import random
NUM_SAMPLES = 100000000
def inside(p):
    x, y = random.random(), random.random()
    return x*x + y*y < 1

count = sc.parallelize(range(0, NUM_SAMPLES)).filter(inside).count()
pi = 4 * count / NUM_SAMPLES
print("Pi is roughly", pi)
```

Summary

- We just created our own PySpark instance with one of the following: local installation using Anaconda, Docker-based installation, or Databrick community edition.

- We know the steps to access local files in the Notebook environment.

- We created a PySpark dataframe using a *movies.csv* file and printed the output.

- We ran a small PySpark command to calculate the value of Pi.

Great job! In the next chapter, we will delve more into PySpark operations and take you through some interesting topics like feature engineering and visualization. Keep learning and stay tuned.

CHAPTER 2

PySpark Basics

This chapter will help you understand the basic operations of PySpark. You are encouraged to set up the PySpark environment and try the following operations on any dataset of your choice for enhanced understanding. Since Spark itself is a very big topic, we will give you just enough content to get you started with PySpark basics and concepts before jumping into data-wrangling activities. This chapter will demonstrate the most common data operations in PySpark that you may encounter in your day-to-day work.

In this chapter, we will cover the following topics:

- Background of PySpark

- PySpark Resilient Distributed Dataset (RDD) and DataFrames

- Data manipulations

PySpark Background

Before leaping to PySpark, let's learn what Spark is and what its advantages are. Spark is an engine for processing large amounts of data. It is written in Scala and runs on a Java virtual machine. A detailed representation of the Apache Spark ecosystem is presented in Figure 2-1.

The Spark core is the underlying general execution engine for the platform. Spark has built-in components for SQL, streaming, machine learning, and GraphX (graph processing).

© Ramcharan Kakarla, Sundar Krishnan and Sridhar Alla 2021
R. Kakarla et al., *Applied Data Science Using PySpark*, https://doi.org/10.1007/978-1-4842-6500-0_2

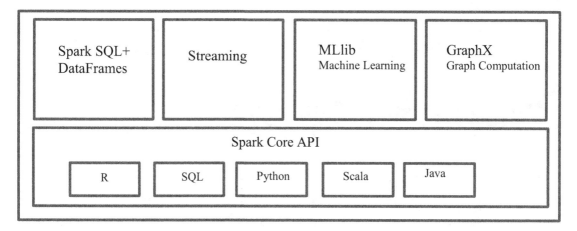

Figure 2-1. *Overview of Spark ecosystem*

- **Spark SQL** is best suited for working with structured data. It enables you to run hive queries at blazing speeds up to 100x faster with the existing data.

- **Streaming** supports nontraditional real-time datasets. It enables interactivity and analytical operations on streaming data, and integrates well with popular data sources, including HDFS, Flume, Kafka, and Twitter.

- **Machine learning** is a scalable library built on top of spark that enables to run different algorithms for specific use cases. It has most of the widely used algorithms off the shelf.

- **GraphX** is a graph computation engine built on top of Spark that enables you to interactively build, transform, and reason about graph-structured data at scale.

Speed is synonymous with Spark and is one of the key drivers of its recent popularity. Spark at its core achieves this via parallel processing and in-memory computations. How does Spark achieve this parallel processing? Well, the answer is Scala, Spark's native language. Scala is functional based, which fits well with parallel processing. Its ease of use for data transformations on large datasets makes it a preferred tool. Spark also has a rich set of libraries, including support for machine learning, SQL, and streaming, making it one of the most versatile tools out there for taming big data in a unified environment.

Now, let's take a step back and see how Spark runs on a cluster to achieve parallel processing. Spark derives its basics from MapReduce. MapReduce is a popular framework for processing large datasets. It consists of two major stages. Map stage is where all the tasks are split among different computers. Reduce stage is where all the keys are shuffled logically and further reduced for calculating meaningful aggregations or doing data transformations. Figure 2-2 illustrates the working of Spark in cluster mode.

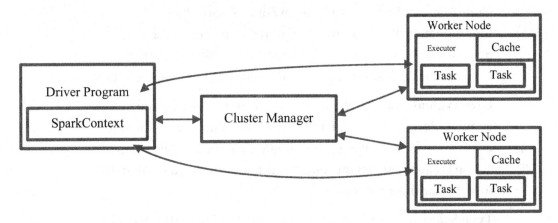

Figure 2-2. *Overview of Spark cluster mode*

When Spark is initiated, an independent process is spun up and is coordinated by SparkContext or Spark Session, as indicated on the left-hand side of the diagram. Cluster Manager assigns the work to the workers. Two workers are shown in Figure 2-2. In practice, there could be *n* number of workers. The task executes a unit of work to the dataset and outputs a new partition dataset. As multiple operations are executed to the data, they benefit from the in-memory computation powered by caching data across these operations. Depending on the type of operation, aggregation results are either sent back to the driver or can be saved onto the disk.

A Directed Acyclic Graph (DAG) is an operator graph that is generated when we submit any code to the Spark console. When action is triggered (either aggregation or data save), Spark submits the graph to the DAG scheduler. The scheduler then divides the operator graphs into stages. Every step can contain several partitions of data, and the DAG scheduler bundles all these individual operator graphs together. The task of workers is to execute these tasks on the executors.

Let's review some concepts and properties of Spark before hopping to RDDs and DataFrames. You may come across these terms in different sections in the book.

Immutable: As the name suggests, something that is immutable can be thought of as a unit of data that cannot be reversed after a transformation. This is extremely important for achieving consistency in computations, especially in a parallel computation setting.

Fault Tolerant: In the case of worker failure during data transformation, the data can be recovered by replaying the transformation on that partition in lineage to achieve the same computation. This eliminates data replication across multiple nodes.

Lazy Evaluation: All transformations in Spark are lazy, meaning computations are not done right away. Instead, Spark remembers all the transformation logic and they are computed only when an action is called. An action could be either returning a result to the driver or saving a dataset.

Type Safety: This indicates that the compiler knows the expected columns and their datatypes.

Data Serialization: This is a method of translating data structures to a specific format to effectively manage the source and destination. By default, Apache Spark uses Java serialization.

Data Locality: Processing tasks are optimized by placing the execution code close to the processed data.

Predicate Pushdown: This technique allows Spark to process only the data that is required by using filter conditions. Only required files and partitions are accessed, thus reducing the disk I/O.

Accumulators: These are global variables assigned to the executors that can be added through the associative and commutative properties.

Catalyst Optimizer: This is an extensible optimizer used to implement Spark SQL and supports both rule-based and cost-based optimization.

So now you may be wondering where PySpark sits in this ecosystem. PySpark can be conceived as a Python-based wrapper on top of the Scala API. PySpark communicates with Spark's Scala-based API through the Py4J library. This library allows Python to interact with the Java virtual machine code. In the next section, we'll look at the ways in which we can interact with data in PySpark via RDDs and DataFrames. We will focus more on the DataFrames, and you will gain deeper insight into why we made that decision in the next section.

PySpark Resilient Distributed Datasets (RDDs) and DataFrames

An *RDD* is an immutable distributed collection of elements of your data. It is partitioned across nodes in your cluster, which can interact with a low-level API with parallel processing. RDDs are flexible in handling structured, semi-structured, and unstructured data. They are best suited for semi- and unstructured data. They can be used to access these data elements with ease and control for any low-level transformation. RDDs are part of Spark's core libraries and only need the Spark context for any of their operations.

RDDs do come with their fair share of challenges. Optimizations and performance gains available for DataFrames are absent for RDDs. If you are working with structured data, the go-to should always be DataFrames. You can always change an RDD to a DataFrame as long as you can define a schema.

A *dataset* is a distributed collection of data. It has benefits of RDDs (strong typing, ability to use powerful lambda functions) with the benefits of Spark SQL's optimized execution engine. A dataset can be constructed from JVM objects and then manipulated using functional transformations (map, flatMap, filter, etc.). The Dataset API is available only in Scala and Java. Python does not have support for the Dataset API. But due to Python's dynamic nature, many of the benefits of the Dataset API are already available (i.e., you can access the field of a row by name naturally: `row.columnName`). The case for R is similar.

A *DataFrame* is a dataset ordered into columns. It holds data in column and row format, like a table. You can perceive the columns to be the variables and the rows the associated data points. It is conceptually equivalent to a table in a relational database or a data frame in R/Python, but with richer optimizations under the hood. DataFrames can be constructed from a wide array of sources, such as structured data files, tables in Hive, external databases, or existing RDDs. The DataFrame API is available in Scala, Java, Python, and R. SQLContext is needed to program DataFrames, as they lie in the SparkSQL area of the Spark ecosystem. You can always convert a DataFrame to an RDD if needed.

Table 2-1 captures the differences between RDDs, DataFrames, and datasets in a nutshell. Remember: We will be using DataFrames as the primary way to interact with data in the book.

Table 2-1. *Features by RDD, DataFrame, and Dataset in Spark*

Feature	RDD	DataFrame	Dataset
Immutable	Yes	Yes	Yes
Fault Tolerant	Yes	Yes	Yes
Type-Safe	Yes	No	Yes
Schema	No	Yes	Yes
Optimizer Engine	NA	Catalyst	Catalyst
Execution Optimization	No	Yes	Yes
API Level Manipulation	Low Level	High Level	High Level
Language Support	Java, Scala, Python, R	Java, Scala, Python, R	Java, Scala

The Spark SQL `DataType` class is the base class of all datatypes, and they are primarily used in all DataFrames (Table 2-2).

Table 2-2. *Datatypes*

Datatype	Broad Classification
BooleanType	Atomic Type: It is an internal type used to represent everything that is not null, arrays, structs, and maps
BinaryType	
DateType	
StringType	
TimestampType	

(*continued*)

Table 2-2. (*continued*)

Datatype	Broad Classification
ArrayType	Non-atomic Types
MapType	
StructType	
CalendarIntervalType	
NumericType	
ShortType	
IntegerType	
LongType	
FloatType	
DoubleType	
DecimalType	
ByteType	
HiveStringType	
ObjectType	
NullType	

In the previous chapter, we introduced different ways you can install and use Spark. Now, let's get to the interesting part and see how you can perform data operations. For the purpose of this book, we will be using the Docker version of PySpark, running on a single machine. If you have a version of PySpark installed on a distributed system, we encourage using it to unleash the power of parallel computing. There will not be any difference in programming or commands between running your operations on a single stand-alone machine or doing so on a cluster. Keep in mind that you will lose out on the processing speeds if you are using a single machine.

For the purpose of demonstrating data operations, we will be using a dataset from themoviedatabase (TMDB: `https://www.themoviedb.org/`). TMDB is a community-built movie and TV database. We used an API to extract a sample of their data. This data contains information on 44,000 movies released internationally. We have the information from Table 2-3 in the dataset. This dataset is in the form of a csv, which is pipe delimited. This dataset will be shared in the chapter's GitHub page.

Table 2-3. *Metadata*

Variable	Description
belongs_to_collection	Indicates whether movie belongs to a collection; collection is specified if exists
budget	Movie's budget
id	Unique identifier for the movie
original_language	Original language in which movie is produced
original_title	Title of the movie
overview	Summary of the movie
popularity	Popularity index of the movie
production_companies	List of companies that produced the movie
production_countries	Country where the movie was produced
release_date	Movie release date
revenue	Movie collection, missing is represented by 0

(*continued*)

Table 2-3. (*continued*)

Variable	Description
runtime	Movie runtime in minutes
status	Indicates whether movie is released or not
tagline	Movie tagline
title	Movie alias English title
vote_average	Average vote rating by the viewers

Data Manipulations

We will work with this data through the end of this chapter as we demonstrate data manipulations.

Once you have set up your connection to Spark using any of the methods demonstrated in the previous chapter, initiate the PySpark session, which can be done using the following code:

```
from pyspark.sql import SparkSession
spark=SparkSession.builder.appName("Data_Wrangling").getOrCreate()
```

Note   The preceding step is universal for any PySpark program.

As we recall from Figure 2-1, SparkSession is the entry point and connects your PySpark code to the Spark cluster. By default, all the configured nodes used for the execution of your code are in a cluster mode setting.

Now, let's read our first dataset. We are using a dataset located on our local machine in the path that we exposed to the Docker session. We are using the Docker method for demonstration purposes. Spark runs on Java 8/11, Scala 2.12, Python 2.7+/3.4+, and R 3.1+. With Java 8 prior to version 8u92, support is deprecated as of Spark 3.0.0. For Python 2 and Python 3 prior to version 3.6, support is deprecated as of Spark 3.0.0. For R prior to version 3.4, support is deprecated as of Spark 3.0.0. Make sure you have the right Java versions for Spark compatibility if you are using a local version. As with Python, all PySpark DataFrame variables are case sensitive.

Reading Data from a File

```
# This is the location where the data file resides
file_location = "movie_data_part1.csv"

# Type of file, PySpark also can read other formats such as json, parquet,
orc
file_type = "csv"

# As the name suggests, it can read the underlying existing schema if
exists
infer_schema = "False"

#You can toggle this option to True or False depending on whether you have
header in your file or not
first_row_is_header = "True"

# This is the delimiter that is in your data file
delimiter = "|"

# Bringing all the options together to read the csv file

df = spark.read.format(file_type) \
.option("inferSchema", infer_schema) \
.option("header", first_row_is_header) \
.option("sep", delimiter) \
.load(file_location)
```

If your dataset is not in a flat file format and exists as a hive table, you can read it easily via a single command as shown next.

Reading Data from Hive Table

Use this code to read data from a hive table:

```
df = spark.sql("select * from database.table_name")
```

Note   Spark only maps the relationship upon reading a data source and will not bring any data into memory.

Executing the preceding commands will make your data accessible to Spark, and you will be all set to experience the magic of PySpark.

Once we have successfully read the data from the source, one of the first questions would be, how can I know the metadata information? One of the handy ways to read metadata is through a simple command.

Reading Metadata

The output is presented in Figure 2-3. We can quickly identify the variables with their datatypes using this function:

df.printSchema()

```
root
 |-- belongs_to_collection: string (nullable = true)
 |-- budget: string (nullable = true)
 |-- id: string (nullable = true)
 |-- original_language: string (nullable = true)
 |-- original_title: string (nullable = true)
 |-- overview: string (nullable = true)
 |-- popularity: string (nullable = true)
 |-- production_companies: string (nullable = true)
 |-- production_countries: string (nullable = true)
 |-- release_date: string (nullable = true)
 |-- revenue: string (nullable = true)
 |-- runtime: string (nullable = true)
 |-- status: string (nullable = true)
 |-- tagline: string (nullable = true)
 |-- title: string (nullable = true)
 |-- vote_average: string (nullable = true)
```

Figure 2-3. *Ouput of printSchema*

There are multiple ways to achieve the preceding output. The same information can also be extracted through using the following command in a Python list format, which is more desirable for further data operations:

df.dtypes

You can also use df.columns to list the columns without the datatype information.

Counting Records

One other quick check you can perform on a dataset is to observe the total number of records. This can be done via the following command:

```
df.count()
```

Output:
43998

You can wrap this in a `print` function in Python to get a more descriptive output, as follows:

```
print('The total number of records in the movie dataset is '+str(df.
count()))
```

Output:
The total number of records in the movie dataset is 43998

We know from previous experience of this dataset that a few of the columns are dictionaries and are not useful in their raw form, and some of the variables contain descriptive information. Let's select a few random variables of our choice and see how they look.

Subset Columns and View a Glimpse of the Data

Again, there are multiple ways to achieve the same output (Figure 2-4) of the following:

```
# Defining a list to subset the required columns
select_columns=['id','budget','popularity','release_
date','revenue','title']

# Subsetting the required columns from the DataFrame
df=df.select(*select_columns)

# The following command displays the data; by default it shows top 20 rows
df.show()
```

```
+------+--------+--------------------+------------+--------+--------------------+
|   id| budget|          popularity|release_date|revenue|               title|
+------+--------+--------------------+------------+--------+--------------------+
|43000|       0|               2.503|  1962-05-23|      0|The Elusive Corporal|
|43001|       0|                5.51|  1962-11-12|      0| Sundays and Cybele|
|43002|       0|                5.62|  1962-05-24|      0|Lonely Are the Brave|
|43003|       0|               7.159|  1975-03-12|      0|          F for Fake|
|43004|  500000|               3.988|  1962-10-09|      0|Long Day's Journe...|
|43006|       0|               3.194|  1962-03-09|      0|          My Geisha|
|43007|       0|               2.689|  1962-10-31|      0|Period of Adjustment|
|43008|       0|               6.537|  1959-03-13|      0|    The Hanging Tree|
|43010|       0|               4.297|  1962-01-01|      0|Sherlock Holmes a...|
|43011|       0|               4.417|  1962-01-01|      0|   Sodom and Gomorrah|
|43012| 7000000|4.7219999999999995|  1962-11-21|4000000|         Taras Bulba|
|43013|       0|               2.543|  1962-04-17|      0|The Counterfeit T...|
|43014|       0|               4.303|  1962-10-24|      0|     Tower of London|
|43015|       0|               3.493|  1962-12-07|      0|Varan the Unbelie...|
|43016|       0|               2.851|  1962-01-01|      0|Waltz of the Tore...|
|43017|       0|               4.047|  1961-10-11|      0|         Back Street|
|43018|       0|               2.661|  1961-06-02|      0|Gidget Goes Hawaiian|
|43019|       0|               3.225|  2010-05-28|      0|Schuks Tshabalala...|
|43020|       0|                5.72|  1961-06-15|      0|The Colossus of R...|
|43021|       0|               3.292|  2008-08-22|      0|          Sex Galaxy|
+------+--------+--------------------+------------+--------+--------------------+
only showing top 20 rows
```

Figure 2-4. *Output of df.show()*

All the operations are done in a single statement, as follows:

```
df.select('id','budget','popularity','release_date','revenue','title').
show()
```

You also have the option of selecting the columns by index instead of selecting the names from the original DataFrame:

```
df.select(df[2],df[1],df[6],df[9],df[10],df[14]).show()
```

Be cautious while using the preceding option if you already used a first method to subset. It may throw an *index out of range* error. This is because the new DataFrame will only contain a subset of columns. You can also change the number of rows to display in the show option to a required number by using the following statement. In this example, we set the number to 25, but it can be changed to any *n*.

```
df.show(25,False)
```

Note   Using False not only allows you to override the default setting of 20 rows, but also displays the truncated content, if any.

Missing Values

You can calculate the missing values in a single column or in multiple columns by using the built-in functions in PySpark, as follows:

```
from pyspark.sql.functions import *
df.filter((df['popularity']=='')|df['popularity'].isNull()|isnan(df['popula
rity'])).count()
```

Output:

215

Let's break down the preceding command and see what is happening. First, we use the `filter` function on the DataFrame and pass multiple conditions based on an OR condition, which is represented by "|" in the expression. In the first condition, we search for a null string present in the *popularity* column. In the second condition, we use the `.isNull()` operator, which returns true when popularity is null; null represents no value or nothing. In the third condition, we use `isnan` to identify NaN (not a number). These are usually a result of mathematical operations that are not defined.

If you need to calculate all the missing values in the DataFrame, you can use the following command:

```
df.select([count(when((col(c)=='') | col(c).isNull() |isnan(c), c)).
alias(c) for c in df.columns]).show()
```

This command selects all the columns and runs the preceding missing checks in a loop. Then when condition is used here to subset the rows that meet the missing value criteria (Figure 2-5).

```
+---+------+----------+------------+-------+-----+
| id|budget|popularity|release_date|revenue|title|
+---+------+----------+------------+-------+-----+
|125|   125|       215|         221|    215|  304|
+---+------+----------+------------+-------+-----+
```

Figure 2-5. *Ouput of missing values across all columns of a DataFrame*

Note   Make sure to import `pyspark.sql`.functions before you run any sort operations.

One-Way Frequencies

Okay, let's see how we can calculate the frequencies of categorical variables (Figure 2-6). Terminology alert: Categorical variables are any string variables that exist in a dataset. Let's verify if there are any repetitive titles in the dataset we have, using the following code:

```
df.groupBy(df['title']).count().show()
```

```
+--------------------+-----+
|               title|count|
+--------------------+-----+
|   The Corn Is Green|    1|
|Meet The Browns -...|    1|
|Morenita, El Esca...|    1|
|  Father Takes a Wife|   1|
|The Werewolf of W...|    1|
|My Wife Is a Gang...|    1|
|Depeche Mode: Tou...|    1|
|   A Woman Is a Woman|   1|
|History Is Made a...|    1|
|      Colombian Love|    1|
|        Ace Attorney|    1|
|     Not Like Others|    1|
|40 Guns to Apache...|    1|
|         Middle Men|    1|
|         It's a Gift|    1|
|    La Vie de Bohème|    1|
|Rasputin: The Mad...|    1|
|The Ballad of Jac...|    1|
|        How to Deal|    1|
|             Freaked|    1|
+--------------------+-----+
only showing top 20 rows
```

Figure 2-6. *Output of one-way frequency*

The preceding command will give you the number of times a title has appeared in a dataset. Often, we would like to see sorted data, so we would use the following (Figure 2-7):

```
df.groupby(df['title']).count().sort(desc("count")).show(10, False)
```

```
+--------------------+-----+
|title               |count|
+--------------------+-----+
|null                |304  |
|Les Misérables      |8    |
|The Three Musketeers|8    |
|Cinderella          |8    |
|The Island          |7    |
|A Christmas Carol   |7    |
|Hamlet              |7    |
|Dracula             |7    |
|Frankenstein        |7    |
|The Lost World      |6    |
+--------------------+-----+
only showing top 10 rows
```

Figure 2-7. *Output of one-way frequency sorted*

Note   groupby is an alias for groupBy; both the variants will work.

There are quite a few titles that are repeated. It looks like having a missing title is the most common issue. Real-world data can be messy. This is good, but we may want to fine-tune the results by eliminating any missing values and adding an additional filter to limit it to only titles that occurred more than four times.

Sorting and Filtering One-Way Frequencies

Let's first filter the values that are not null. We used the not-equal sign (!= and ~) to create a temporary dataset, which is done intentionally to demonstrate the use of the not condition.

```
# Subsetting and creating a temporary DataFrame to eliminate any missing
values

df_temp=df.filter((df['title']!='')&(df['title'].isNotNull()) &
(~isnan(df['title'])))

# Subsetting the DataFrame to titles that are repeated more than four times

df_temp.groupby(df_temp['title']).count().filter("`count` >4").
sort(col("count").desc()).show(10,False)
```

If you don't specify the desc function, the results will be sorted in ascending order by default (Figure 2-8).

```
+--------------------+-----+
|title               |count|
+--------------------+-----+
|Les Misérables      |8    |
|The Three Musketeers|8    |
|Cinderella          |8    |
|Hamlet              |7    |
|The Island          |7    |
|Dracula             |7    |
|A Christmas Carol   |7    |
|Frankenstein        |7    |
|Framed              |6    |
|The Lost World      |6    |
+--------------------+-----+
only showing top 10 rows
```

Figure 2-8. *Output of filtered and sorted version of one-way frequency*

```
# The following command is to find the number of titles that are repeated
four times or more
```

```
df_temp.groupby(df_temp['title']).count().filter("`count` >=4").
sort(col("count").desc()).count()
```

Output:

43

```
# The following command is to delete any temporary DataFrames that we
created in the process
```

```
del df_temp
```

In the preceding commands, you may have observed that there are multiple ways to achieve the same result. As you become fluent in the language, you may develop your own preferences.

Casting Variables

As you have observed from the metadata, all the variables are indicated as strings. After observing the data, we know some of these fields are either integers, floats, or date objects. Some of the operations can lead to misleading results if the datatypes are not correctly identified. We strongly recommend you do the due diligence of identifying the right datatype for any of your analyses. We will now see how we can cast some of these variables to the right data types.

If you are not careful in identifying the datatypes, you may experience data loss on casting. For example, if you are converting a string column to a numeric one, the resulting column can be all nulls. Try this with the column *tagline* in the demo dataset and observe the result. It is good practice to explore the dataset before applying any transformations. Some typos in the manual data (e.g., 12,1 instead of 12.1) can also result in unexpected results.

Let's first print the datatypes before casting, convert the budget variable to float, and observe the change (Figures 2-9 and 2-10).

```
#Before Casting
df.dtypes
```

```
[('id', 'string'),
 ('budget', 'string'),
 ('popularity', 'string'),
 ('release_date', 'string'),
 ('revenue', 'string'),
 ('title', 'string')]
```

Figure 2-9. *Datatypes before casting*

```
#Casting
df = df.withColumn('budget',df['budget'].cast("float"))
```

```
#After Casting
df.dtypes
```

```
[('id', 'string'),
 ('budget', 'float'),
 ('popularity', 'string'),
 ('release_date', 'string'),
 ('revenue', 'string'),
 ('title', 'string')]
```

Figure 2-10. *Datatypes after casting*

The cast function is used. Also note we used the additional function .withColumn. This is one of the most common functions used in PySpark. It is used for updating the values of, renaming, and converting datatypes, and for creating new columns (Figures 2-11 and 2-12). You will have a better appreciation of its use further down the lane. You may be wondering how you can extend the function to multiple columns. If you already know your datatypes, you can use the lists of identified variables and iterate them over a loop.

```
#Importing necessary libraries
from pyspark.sql.types import *
```

```
#Identifying and assigning lists of variables
int_vars=['id']
float_vars=['budget', 'popularity', 'revenue']
date_vars=['release_date']
```

```
#Converting integer variables
```

```
for column in int_vars:
    df=df.withColumn(column,df[column].cast(IntegerType()))
for column in float_vars:
    df=df.withColumn(column,df[column].cast(FloatType()))
for column in date_vars:
    df=df.withColumn(column,df[column].cast(DateType()))

df.dtypes
```

```
[('id', 'int'),
 ('budget', 'float'),
 ('popularity', 'float'),
 ('release_date', 'date'),
 ('revenue', 'float'),
 ('title', 'string')]
```

Figure 2-11. *Datatypes of all variables after casting*

```
df.show(10,False)
```

```
+-----+--------+----------+------------+-------+-------------------------------------+
|id   |budget  |popularity|release_date|revenue|title                                |
+-----+--------+----------+------------+-------+-------------------------------------+
|43000|0.0     |2.503     |1962-05-23  |0.0    |The Elusive Corporal                 |
|43001|0.0     |5.51      |1962-11-12  |0.0    |Sundays and Cybele                   |
|43002|0.0     |5.62      |1962-05-24  |0.0    |Lonely Are the Brave                 |
|43003|0.0     |7.159     |1975-03-12  |0.0    |F for Fake                           |
|43004|500000.0|3.988     |1962-10-09  |0.0    |Long Day's Journey Into Night        |
|43006|0.0     |3.194     |1962-03-09  |0.0    |My Geisha                            |
|43007|0.0     |2.689     |1962-10-31  |0.0    |Period of Adjustment                 |
|43008|0.0     |6.537     |1959-03-13  |0.0    |The Hanging Tree                     |
|43010|0.0     |4.297     |1962-01-01  |0.0    |Sherlock Holmes and the Deadly Necklace|
|43011|0.0     |4.417     |1962-01-01  |0.0    |Sodom and Gomorrah                   |
+-----+--------+----------+------------+-------+-------------------------------------+
only showing top 10 rows
```

Figure 2-12. *Glimpse of data after changing the datatype*

In the preceding output, if you observe closely, there is a visible difference in how the values of the *budget* and *revenue* columns are displayed with a ".0" value, indicating the datatype. You can compare Figure 2-4 with Figure 2-12 to observe these differences before and after casting.

Descriptive Statistics

To analyze any data, you should have a keen understanding of the type of data, its distribution, and its dispersion. Spark has a nice suite of built-in functions that will make it easier to quickly calculate these fields. The `describe` function in Spark is very handy, as it gives the count of total non-missing values for each column, mean/average, standard deviation, and minimum and maximum values (Figure 2-13). You can transform the output of this DataFrame and use it later in missing-value imputations. We will discuss this in more depth in the model-building section.

```
df.describe()
```

```
+-------+-------------------+--------------------+------------------+--------------------+--------------------+
|summary|                 id|              budget|        popularity|             revenue|               title|
+-------+-------------------+--------------------+------------------+--------------------+--------------------+
|  count|              43784|               43873|             43783|               43783|               43694|
|   mean|  44502.304312077475|    3736901.834963166| 5.295444259579189|   9697079.597382545|            Infinity|
| stddev| 27189.646588626343|1.5871814952777334E7|6.168030519208248|5.6879384496288106E7|                 NaN|
|    min|                  2|                 0.0|               0.6|                 0.0|!Women Art Revolu...|
|    max|             100988|               3.8E8|             180.0|        2.78796518E9|       시크릿 Secret|
+-------+-------------------+--------------------+------------------+--------------------+--------------------+
```

Figure 2-13. *Describe function output*

Mean presented here is arithmetic mean, which is the central value of a discrete set of observations. Mean is a good estimate for many operations but can be susceptible to the influence of outliers. Outliers can have undesirable effects on the mean value. *Median* is a better estimate as it is not influenced by outliers, but it is expensive to calculate. It is the same with Spark. This operation is expensive because of sorting. To make this calculation faster, Spark implemented a variant of the Greenwald-Khanna algorithm with some speed optimizations. The function used for calculating the median is `approxQuantile`.

Three parameters have to be passed through `approxQuantile` function, as follows:

- col – the name of the numerical column

- probabilities – a list of quantile probabilities. Each number must belong to [0, 1]. For example, 0 is the minimum, 0.5 is the median, 1 is the maximum.

- relativeError – The relative target precision to achieve (>= 0). If set to 0, the exact quantiles are computed, which could be very expensive. Note that values greater than 1 are accepted but give the same result as 1.

Let's try calculating the median for the *budget* column.

#Since unknown values in budget are marked to be 0, let's filter out those values before calculating the median

```
df_temp = df.filter((df['budget']!=0)&(df['budget'].isNotNull()) &
(~isnan(df['budget'])))
```

#Here the second parameter indicates the median value, which is 0.5; you can also try adjusting the value to calculate other percentiles

```
median=df.approxQuantile('budget',[0.5],0.1)
```

#Printing the Value
```
print ('The median of budget is '+str(median))
```

Output:

```
The median of budget is [7000000.0]
```

If you wanted to calculate multiple columns' medians at once, you would just need to change the first parameter to `list`.

Unique/Distinct Values and Counts

You may sometimes just want to know the number of levels (cardinality) within a variable. You can do this using the `countDistinct` function available in Spark (Figure 2-14).

```
# Counts the distinct occurances of titles
df.agg(countDistinct(col("title")).alias("count")).show()
```

```
+-----+
|count|
+-----+
|41138|
+-----+
```

Figure 2-14. *Distinct titles count output*

Alternatively, you can also look into the distinct occurrences using the `distinct` function (Figure 2-15).

```
# Counts the distinct occurances of titles
df.select('title').distinct().show(10,False)
```

```
+----------------------------------------------+
|title                                         |
+----------------------------------------------+
|The Corn Is Green                             |
|Meet The Browns - The Play                    |
|Morenita, El Escandalo                        |
|Father Takes a Wife                           |
|The Werewolf of Washington                    |
|My Wife Is a Gangster                         |
|Depeche Mode: Touring the Angel Live in Milan |
|A Woman Is a Woman                            |
|History Is Made at Night                      |
|Colombian Love                                |
+----------------------------------------------+
only showing top 10 rows
```

Figure 2-15. *Distinct titles output*

Intuitively, you may also want to check out distinct titles by year. So, let's first extract the year from the release date.

```
# Extracting year from the release date
df_temp=df.withColumn('release_year',year('release_date'))
```

You should be able to extract month and day from the date variable using similar functions if needed. The following two commands are shown for demonstration purposes and are not required for our desired output. The third command gets you the distinct counts by year (Figure 2-16).

```
# Extracting month
df_temp=df_temp.withColumn('release_month',month('release_date'))
```

```
# Extracting day of month
df_temp=df_temp.withColumn('release_day',dayofmonth('release_date'))
```

```
# Calculating the distinct counts by the year
df_temp.groupBy("release_year").agg(countDistinct("title")).show(10,False)
```

```
+------------+------------+
|release_year|count(title)|
+------------+------------+
|1959        |271         |
|1990        |496         |
|1975        |365         |
|1977        |415         |
|1924        |19          |
|2003        |1199        |
|2007        |1896        |
|2018        |4           |
|1974        |434         |
|2015        |13          |
+------------+------------+
only showing top 10 rows
```

Figure 2-16. *Distinct titles count by year*

Again, you should be able to sort the output either by descending or ascending order using the functions we discussed in the sorting section.

Filtering

Spark offers multiple ways to filter data. In fact, we encountered the `filter` function even before getting into this section. It demonstrates how important this function is even in basic operations. You may also encounter the `where` function for filtering. Both these functions work the same way. However, `filter` is the standard Scala name for such a function, and `where` is for people who prefer SQL.

```
.where() = .filter()
```

Earlier, we encountered conditions in `filter` that included or and and conditions. In this demonstration, let's try other available methods to filter the data. Say I would like to filter all the titles that start with "Meet"; how should we approach it? If you are a SQL user, you can do it simply by using a `like` condition, as shown here. PySpark offers regex functions that can be used for this purpose (Figure 2-17).

```
df.filter(df['title'].like('Meet%')).show(10,False)
```

```
+-----+----------+----------+------------+-----------+--------------------------------+
|id   |budget    |popularity|release_date|revenue    |title                           |
+-----+----------+----------+------------+-----------+--------------------------------+
|43957|500000.0  |2.649     |2005-06-28  |1000000.0  |Meet The Browns - The Play|
|39997|0.0       |3.585     |1989-11-15  |0.0        |Meet the Hollowheads            |
|16710|0.0       |11.495    |2008-03-21  |4.1939392E7|Meet the Browns                 |
|20430|0.0       |3.614     |2004-01-29  |0.0        |Meet Market                     |
|76435|0.0       |1.775     |2011-03-31  |0.0        |Meet the In-Laws                |
|76516|5000000.0 |4.05      |1990-11-08  |485772.0   |Meet the Applegates             |
|7278 |3.0E7     |11.116    |2008-01-24  |8.4646832E7|Meet the Spartans               |
|32574|0.0       |7.42      |1941-03-14  |0.0        |Meet John Doe                   |
|40506|0.0       |4.814     |1997-01-31  |0.0        |Meet Wally Sparks               |
|40688|2.4E7     |6.848     |1998-03-27  |4562146.0  |Meet the Deedles                |
+-----+----------+----------+------------+-----------+--------------------------------+
only showing top 10 rows
```

Figure 2-17. *Filtering matches using "like" expression*

Now, let's find out the titles that *do not* end with an "s" (Figure 2-18).

```
df.filter(~df['title'].like('%s')).show(10,False)
```

```
+-----+----------+----------+------------+-------+-----------------------------------------+
|id   |budget    |popularity|release_date|revenue|title                                    |
+-----+----------+----------+------------+-------+-----------------------------------------+
|43000|0.0       |2.503     |1962-05-23  |0.0    |The Elusive Corporal                     |
|43001|0.0       |5.51      |1962-11-12  |0.0    |Sundays and Cybele                       |
|43002|0.0       |5.62      |1962-05-24  |0.0    |Lonely Are the Brave                     |
|43003|0.0       |7.159     |1975-03-12  |0.0    |F for Fake                               |
|43004|500000.0  |3.988     |1962-10-09  |0.0    |Long Day's Journey Into Night            |
|43006|0.0       |3.194     |1962-03-09  |0.0    |My Geisha                                |
|43007|0.0       |2.689     |1962-10-31  |0.0    |Period of Adjustment                     |
|43008|0.0       |6.537     |1959-03-13  |0.0    |The Hanging Tree                         |
|43010|0.0       |4.297     |1962-01-01  |0.0    |Sherlock Holmes and the Deadly Necklace|
|43011|0.0       |4.417     |1962-01-01  |0.0    |Sodom and Gomorrah                       |
+-----+----------+----------+------------+-------+-----------------------------------------+
only showing top 10 rows
```

Figure 2-18. *Filtering non-matches using "like" expression*

If we wanted to find any title that contains "ove," we could use the `rlike` function, which is a regular expression (regex; Figure 2-19).

```
df.filter(df['title'].rlike('\w*ove')).show(10,False)
```

```
+-----+------+----------+------------+------------+-----------------------+
|id   |budget|popularity|release_date|revenue     |title                  |
+-----+------+----------+------------+------------+-----------------------+
|43100|0.0   |7.252     |1959-10-07  |0.0         |General Della Rovere   |
|43152|0.0   |5.126     |2001-06-21  |0.0         |Love on a Diet         |
|43191|0.0   |4.921     |1952-08-29  |0.0         |Beware, My Lovely      |
|43281|0.0   |2.411     |1989-11-22  |0.0         |Love Without Pity      |
|43343|0.0   |3.174     |1953-12-25  |0.0         |Easy to Love           |
|43347|3.0E7 |14.863    |2010-11-22  |1.02820008E8|Love & Other Drugs     |
|43362|0.0   |1.705     |1952-02-23  |0.0         |Love Is Better Than Ever|
|43363|0.0   |2.02      |1952-05-29  |0.0         |Lovely to Look At      |
|43395|0.0   |4.758     |1950-11-10  |0.0         |Two Weeks with Love    |
|43455|0.0   |4.669     |1948-08-23  |0.0         |The Loves of Carmen    |
+-----+------+----------+------------+------------+-----------------------+
only showing top 10 rows
```

Figure 2-19. *Filtering matches using regular expression*

The preceding expression can also be rewritten as follows:

```
df.filter(df.title.contains('ove')).show()
```

There are situations where we have thousands of columns and want to identify or subset the columns by a particular prefix or suffix. We can achieve this using the colRegex function. First, let's identify the variables that start with "re" (Figure 2-20).

```
df.select(df.colRegex("`re\w*`")).printSchema()
```

```
root
 |-- release_date: date (nullable = true)
 |-- revenue: float (nullable = true)
```

Figure 2-20. *Filtering columns by prefix using regular expressions*

Okay, so how can we identify variables that end with a particular suffix? Here we have three variables that end with "e"; let's see if we can identify them by tweaking our earlier regular expressions (Figure 2-21).

```
df.select(df.colRegex("`\w*e`")).printSchema()
```

```
root
 |-- release_date: date (nullable = true)
 |-- revenue: float (nullable = true)
 |-- title: string (nullable = true)
```

Figure 2-21. *Filtering columns by suffix using regular expressions*

53

Regex can be confusing, so let's have a brief refresher on the topic. In general, we can identify all the characters and numerics by metacharacters. Regex uses these metacharacters to match any literal or alphanumeric expression. Here is a short list of the most-used metacharacters:

Single characters:

- \d – Identifies numbers between 0 and 9

- \w – Identifies all upper- and lowercase alphabets and numbers from 0 to 9 [A–Z a–z 0–9]

- \s – Whitespace

- . – Any character

We also have a set of quantifiers that guide how many characters to search, as follows:

Quantifiers:

- * – 0 or more characters

- + – 1 or more characters

- ? – 0 or 1 characters

- {min,max} – specify the range, any characters between the range including min, max

- {n} – Exactly *n* characters

If you look back at both the earlier expressions, you will see that we have made use of \w, indicating we want to match all alphanumeric characters, ignoring the case. In the expression where we wanted to identify variables with the "re" prefix, \w was followed by the quantifier * to indicate we wanted to include all characters after "re."

Creating New Columns

Creating new features (columns) plays a critical role in many analytical applications. Spark offers multiple ways in which you can create new columns. The simplest method is through the `withColumn` function. Let's say we want to calculate the variance of popularity. Variance is a measure of how far the numbers are spread from the mean and is represented by following formula:

$$S^2 = \sum(X_i - \bar{X})^2 / (n-1)$$

x_i – Individual observations

$\bar{x}$ – Mean

n – Total number of observations

So, let's first calculate the mean by using the following command. The agg function used here is handy instead of using describe when you are looking for a specific statistic:

```
mean_pop=df.agg({'popularity': 'mean'}).collect()[0]['avg(popularity)']
count_obs= df.count()
```

Let's add this to all the rows since we need this value along with individual observation values using the withColumn and lit functions. The lit function is a way to interact with column literals. It is very useful when you want to create a column with a value directly.

```
df=df.withColumn('mean_popularity',lit(mean_pop))
```

The *mean_popularity* column has the same value for all the rows. The pow function in the following command helps you the raise number to a power. Here, we would need to square the number, hence 2 is passed as the parameter.

```
df=df.withColumn('varaiance',pow((df['popularity']-df['mean_
popularity']),2))
```

```
variance_sum=df.agg({'varaiance': 'sum'}).collect()[0]['sum(varaiance)']
```

The preceding command sums up all the differences across all the rows. And in the final step, we divide the sum by the number of observations, which yields us the result.

```
variance_population= variance_sum/(count_obs-1)
```

Output:

```
37.85868805766277
```

Although this variance calculation involved multiple steps, the idea is to give you a flavor of operations that might be involved in creating a new feature. Once you have the value of a variance you can create custom variables based on it.

Let's step back. You may ask, "If I have multiple variables to create, how can I do it all at once and keep them well structured?" We will walk you through the solution next. Say I would like to create variables that can classify both the budget and the popularity variables into high, medium, and low. First, let's define what the high, medium, and low thresholds are based on. The following thresholds are used for convenience; you can alter them to values based on more rigorous analysis.

All we are doing in the following piece of code is defining a simple Python function. We are passing two variables into the function, and based on threshold we decide we are returning the new variables, which are both strings.

```
def new_cols(budget,popularity):
 if budget<10000000: budget_cat='Small'
 elif budget<100000000: budget_cat='Medium'
 else: budget_cat='Big'
 if popularity<3: ratings='Low'
 elif popularity<5: ratings='Mid'
 else: ratings='High'
 return budget_cat,ratings
```

Now, here is the fun part: we are now using user-defined functions. So, what are user-defined functions? If you are coming from a Python world, these are very similar to the .map() and .apply()methods for the Pandas DataFrame and series. In simple words, user-defined functions allow us to use values from rows in a DataFrame as input and can map it to the entire DataFrame. One important fact here is that you will have to specify the output datatype. Unfortunately, there is no getting around this. This can be painful when you have to write out multiple columns. We will discuss another method after this that can be used as an alternative.

In the command, we define our user-defined function. Observe that we create StructType to hold the new variables. We also explicitly define what the datatype is that will be returned. We know both the return types are strings, so we use StringType(). StructType objects define the schema of Spark DataFrames. StructType objects contain a list of StructField objects that define the name, type, and nullable flag for each column in a DataFrame.

StructType Objects are really useful for eliminating any order dependencies in your code. What do we mean by that? Well, if you have multiple functions that need to run before the final step, you can wrap them all under the StructType so as to return an array of results rather than create multiple columns. This is useful for reading and managing long codebases that involve multiple transformations and actions. For this example, we will split the StructType that holds multiple outputs into individual columns for any further access.

```
# Apply the user-defined function on the DataFrame
udfB=udf(new_cols,StructType([StructField("budget_cat", StringType(),
True),StructField("ratings", StringType(), True)]))
```

In the following command, we pass a user-defined function with two input columns: *budget* and *popularity*:

```
temp_df=df.select('id','budget','popularity').withColumn("newcat",udfB("bud
get","popularity"))
```

```
# Unbundle the struct type columns into individual columns and drop the
struct type
df_with_newcols = temp_df.select('id','budget','popularity','newc
at').withColumn('budget_cat', temp_df.newcat.getItem('budget_cat')).
withColumn('ratings', temp_df.newcat.getItem('ratings')).drop('newcat')
df_with_newcols.show(15,False)
```

In the following command, note that we are again using the withColumn function to extract data inside the StructType. Since we have the StructType holding string datatypes, we use the getItem function to access the elements within the StructType. Observe we are concatenating the functions one after the other to achieve the desired output in a single line (Figure 2-22).

```
+-----+---------+----------+----------+-------+
|id   |budget   |popularity|budget_cat|ratings|
+-----+---------+----------+----------+-------+
|43000|0.0      |2.503     |Small     |Low    |
|43001|0.0      |5.51      |Small     |High   |
|43002|0.0      |5.62      |Small     |High   |
|43003|0.0      |7.159     |Small     |High   |
|43004|500000.0 |3.988     |Small     |Mid    |
|43006|0.0      |3.194     |Small     |Mid    |
|43007|0.0      |2.689     |Small     |Low    |
|43008|0.0      |6.537     |Small     |High   |
|43010|0.0      |4.297     |Small     |Mid    |
|43011|0.0      |4.417     |Small     |Mid    |
|43012|7000000.0|4.722     |Small     |Mid    |
|43013|0.0      |2.543     |Small     |Low    |
|43014|0.0      |4.303     |Small     |Mid    |
|43015|0.0      |3.493     |Small     |Mid    |
|43016|0.0      |2.851     |Small     |Low    |
+-----+---------+----------+----------+-------+
only showing top 15 rows
```

Figure 2-22. *Output of newly created DataFrame with custom columns*

Another way we can achieve the same result is through the when function. One advantage of using this function is you don't have to define the output data type. This is handy for quick and dirty operations. Let's recreate the preceding columns using the when function.

```
df_with_newcols = df.select('id','budget','popularity').\
withColumn('budget_cat', when(df['budget']<10000000,'Small').when(df['budge
t']<100000000,'Medium').otherwise('Big')).\
withColumn('ratings', when(df['popularity']<3,'Low').
when(df['popularity']<5,'Mid').otherwise('High'))
```

If you observe the preceding command, you can see we just concatenated a bunch of when conditions to define our new variables. This produces the same output as the previous way we discussed. Another useful tip: Using a combination of withColumn and when functions you can update the value of an existing column if you are using the same column name. You don't have to create a new temporary DataFrame (as just shown) for the operations. Instead, you can use the same one to update or add column values.

Deleting and Renaming Columns

You can always drop any column or columns using the drop function.

```
columns_to_drop=['budget_cat']
df_with_newcols=df_with_newcols.drop(*columns_to_drop)
```

You can verify the output by using the printSchema function.

Renaming can be done using either the withColumnRenamed function or the alias function.

```
df_with_newcols = df_with_newcols.withColumnRenamed('id','film_id') .withCo
lumnRenamed('ratings','film_ratings')
```

If you would like to change multiple column names, you try the following command:

```
# You can define all the variable changes in the list
new_names = [('budget','film_budget'),('popularity','film_popularity')]
```

```
# Applying the alias function
df_with_newcols_renamed = df_with_newcols.select(list(map(lambda
old,new:col(old).alias(new),*zip(*new_names))))
```

Both the preceding methods will have the same execution plan. You can verify all these outputs by using the `printSchema` function.

Use the following dataset for Exercise 2-1: `https://www.kaggle.com/kakarlaramcharan/tmdb-data-0920`.

EXERCISE 2-1: DATA MANIPULATIONS

Question 1: Identify the titles that were repeated between the years 2000 and 2015 and count number of titles (Hint: Use filter).

Question 2: Identify all titles that contain Harry in the title name.

Question 3: Create a new column as a binary indicator of whether the original language is English.

Question 4: Tabulate the mean of popularity by year.

Summary

- We got an overview of the Spark ecosystem and how PySpark fits in this environment, and also learned the advantages of using Spark.

- We now know the differences between RDDs, DataFrames, and datasets.

- We ventured our way into data manipulations and learned how to use various built-in functions.

- We also defined our own user-defined function and saw its implementation and mechanics.

Good job! You are now familiar with some of the key concepts that will be useful throughout your journey in this book. We will take a deep dive into some of the intermediate and advanced data manipulation techniques in the next chapter.

CHAPTER 3

Utility Functions and Visualizations

In this chapter, we will dive into the utility of some of the advanced functions available in PySpark. You are encouraged to read the previous chapter and try the following operations on any dataset of your choice to improve your understanding. This chapter will focus on the windowing functions and other topics that will be useful in the creation and application of Spark programs on large datasets. We will also introduce the visualization and machine learning processes as we conclude the chapter. This should enable to you step into the next machine learning chapters with ease.

In this chapter, we will cover the following topics:

- Additional data manipulations

- Data visualizations

- Machine learning overview

Additional Data Manipulations

Let's start by reviewing some data manipulations that were not covered in the previous chapter.

© Ramcharan Kakarla, Sundar Krishnan and Sridhar Alla 2021
R. Kakarla et al., *Applied Data Science Using PySpark*, https://doi.org/10.1007/978-1-4842-6500-0_3

String Functions

For the purpose of this section, let's rerun the commands we executed in Chapter 2.

```
# Rerunning the following command for recreating budget_cat variable

df_with_newcols = df.select('id','budget','popularity').\
withColumn('budget_cat', when(df['budget']<10000000,'Small').when(df['budge
t']<100000000,'Medium').otherwise('Big')).\
withColumn('ratings', when(df['popularity']<3,'Low').
when(df['popularity']<5,'Mid').otherwise('High'))
```

If we want to concatenate the values of budget_cat and ratings together into a single column, we can do so using the concat function. On top of this, let's change the case of the new column to lowercase and trim away any white spaces using the lower and trim functions (Figure 3-1).

```
# Concatenating two variables

df_with_newcols=df_with_newcols.withColumn('BudgetRating_
Category',concat(df_with_newcols.budget_cat,df_with_newcols.ratings))

# Changing the new variable to lowercase

df_with_newcols=df_with_newcols.withColumn('BudgetRating_
Category',trim(lower(df_with_newcols.BudgetRating_Category)))

df_with_newcols.show()
```

```
+-----+----------+----------+-----------+-------+---------------------+
|  id |   budget | popularity|budget_cat|ratings|BudgetRating_Category|
+-----+----------+----------+-----------+-------+---------------------+
|43000|      0.0 |    2.503 |     Small |   Low |            smalllow |
|43001|      0.0 |     5.51 |     Small |  High |           smallhigh |
|43002|      0.0 |     5.62 |     Small |  High |           smallhigh |
|43003|      0.0 |    7.159 |     Small |  High |           smallhigh |
|43004| 500000.0 |    3.988 |     Small |   Mid |            smallmid |
|43006|      0.0 |    3.194 |     Small |   Mid |            smallmid |
|43007|      0.0 |    2.689 |     Small |   Low |            smalllow |
|43008|      0.0 |    6.537 |     Small |  High |           smallhigh |
|43010|      0.0 |    4.297 |     Small |   Mid |            smallmid |
|43011|      0.0 |    4.417 |     Small |   Mid |            smallmid |
|43012|7000000.0 |    4.722 |     Small |   Mid |            smallmid |
|43013|      0.0 |    2.543 |     Small |   Low |            smalllow |
|43014|      0.0 |    4.303 |     Small |   Mid |            smallmid |
|43015|      0.0 |    3.493 |     Small |   Mid |            smallmid |
|43016|      0.0 |    2.851 |     Small |   Low |            smalllow |
|43017|      0.0 |    4.047 |     Small |   Mid |            smallmid |
|43018|      0.0 |    2.661 |     Small |   Low |            smalllow |
|43019|      0.0 |    3.225 |     Small |   Mid |            smallmid |
|43020|      0.0 |     5.72 |     Small |  High |           smallhigh |
|43021|      0.0 |    3.292 |     Small |   Mid |            smallmid |
+-----+----------+----------+-----------+-------+---------------------+
only showing top 20 rows
```

Figure 3-1. *Output of string functions*

Registering DataFrames

We have worked so far with DataFrames and their functions. Wouldn't it be great if we could apply the power of SQL functions to these DataFrames? Fortunately, PySpark has a way to access your DataFrames to run SQL queries on top of them. All the operations will be powered by Spark. For this to happen, you will have to register the DataFrame as a temporary table using the `registerTempTable` function. This creates an in-memory table. Note that this table is only available in the Spark session in which it was created. The data is stored in hive's column format. This gives you the ability to access all the columns, just as you do in a regular SQL table. You can rewrite the existing temp table with a different data.

Let's register the DataFrame as a temporary table and see how can we apply simple SQL to achieve the desired results without using any functions (Figure 3-2).

```
# Registering temporary table

df_with_newcols.registerTempTable('temp_data')

# Applying the function to show the results

spark.sql('select ratings, count(ratings) from temp_data group by
ratings').show(False,10)
```

```
+-------+-------------+
|ratings|ratings_count|
+-------+-------------+
|   High|        16856|
|    Low|        14865|
|    Mid|        12277|
+-------+-------------+
```

Figure 3-2. *Output of Spark SQL output on a temp table*

Window Functions

What are window functions? Window functions facilitate the performance of calculations across a set of rows in reference to an existing row. These are extremely useful when you want to calculate analytical functions pertaining to a time. A lot of feature engineering revolves around windowing functions. If you have to classify windowing functions, you can put them in three different boxes. Aggregate functions are those where multiple rows are grouped together to form a single value summary. These functions can be sum, average, min, max, or count, to name a few. Ranking and Analytical are the other two types of window function.

When you are using a window function, you will have to be specific with three important components:

- Partition: This specifies which rows will be included in the same partition with the row. For example, if you want to calculate the top budget movies by the year, you may want to include the year in the partition.

- Order: This specifies how you would like to order the rows inside the partition, either ascending or descending.

- Frame: This specifies how many rows you want to include above/below the current row based on relative position to the current position.

PySpark has a built-in set of functions for ranking and analytical functions. Table 3-1 is a short and most useful functions list.

Table 3-1. *Types of Windowing Functions*

Type	Functions
Ranking	rank
	denseRank
	percentRank
	rowNumber
	ntile
Analytical	lead
	lag
	lastValue
	firstValue
	cumeDist

You will get a better understanding of these concepts with the following illustrations. Let's calculate the deciles of popularity. What do we mean by *deciling*? If you are from an analytical background, you already have an appreciation for deciles. Deciling is a process of dividing the distribution into ten equal buckets in either ascending or descending order. These are used as an input step in evaluating a predictive model's performance. We will take a deeper dive into model performance and validation statistics in later chapters. For now, let's see how we can calculate the deciles for popularity.

```
# Importing the window functions

from pyspark.sql.window import *

# Step 1: Filtering the missing values
df_with_newcols=df_with_newcols.filter( (df_with_newcols['popularity'].
isNotNull()) & (~isnan(df_with_newcols['popularity'])) )

# Step 2: Applying the window functions for calculating deciles
df_with_newcols = df_with_newcols.select("id","budget","popularity",
ntile(10).over(Window.partitionBy().orderBy(df_with_newcols['popularity'].
desc())).alias("decile_rank"))

# Step 3:Dispalying the values
df_with_newcols.groupby("decile_rank").agg(min('popularity').alias('min_
popularity'),max('popularity').alias('max_popularity'),count('populari
ty')).show()
```

Let's break down what's happening in step 2. We use the `ntile` function, which divides the variable distribution into *n* buckets. We also specify how we would like to divide the distribution using a partition. Here, we did not mention any partition for the data. So inherently it will use the entire data. We then use an `orderBy` to order all the rows in the data by `popularity` value in descending order. Finally, the `alias` function is used to rename the `decile` column. Since Spark works on lazy evaluation, the execution of this command will not return the result. At step 3, we are performing an action. As we execute the step, underlying plans are executed and deciles are calculated.

In step 3, we gather all the statistics we are interested in using in the `agg` function from the grouped DataFrame. We also add the minimum, maximum, and count values to the aggregate function. We would expect to see an approximately equal number of records in each decile. We also add the count to cross-verify this fact. The result is shown in Figure 3-3.

```
+-----------+---------------+---------------+-----------------+
|decile_rank|min_popularity |max_popularity |count(popularity)|
+-----------+---------------+---------------+-----------------+
|          1|         10.185|          180.0|             4379|
|          2|          7.481|         10.182|             4379|
|          3|          5.841|          7.481|             4379|
|          4|          4.823|          5.841|             4378|
|          5|          4.054|          4.822|             4378|
|          6|          3.383|          4.054|             4378|
|          7|          2.747|          3.383|             4378|
|          8|          2.075|          2.747|             4378|
|          9|          1.389|          2.075|             4378|
|         10|            0.6|          1.389|             4378|
+-----------+---------------+---------------+-----------------+
```

Figure 3-3. *Output of popularity deciles*

We now have a fair idea of how popularity ratings vary across ten equally distributed buckets. Let's say I want to know the second most popular movie in the year 1970. How can we find it using the functions we just learned?

Before jumping in, let's use our DataFrame in memory. What if I have already overwritten the data? Okay, at this point, if you have overwritten your DataFrame, you can always reload the data. But remember to cast the variables to the right datatypes.

```
# Step 1: Import the window functions

from pyspark.sql.window import *

# Step 2: Select the required subset of columns

df_second_best = df.select('id', 'popularity', 'release_date')

# Step 3: Create the year column from release date

df_second_best=df_second_best.withColumn('release_year',year('release_
date')).drop('release_date')

# Step 4: Define partition function

year_window = Window.partitionBy(df_second_best['release_year']).
orderBy(df_with_newcols['popularity'].desc())
```

```
# Step 5: Apply partition function
```

```
df_second_best=df_second_best.select('id', 'popularity', 'release_
year',rank().over(year_window).alias("rank"))
```

```
# Step 6: Find the second best rating for the year 1970
```

```
df_second_best.filter((df_second_best['release_year']==1970) & (df_second_
best['rank']==2)).show()
```

Let's rewind and observe what's happening in the preceding steps. First, we selected the required columns and extracted the year from the release date. You must be familiar with this step from the earlier section. In step 4, we defined our window function. We want to partition the data by year. The reason for doing this is we want to know the second highest rating in the year 1970. This partition will give us the ability to rank all popularity ratings in the data by year. In step 5, we used a rank function along with the window.partitionBy function we defined earlier to arrive at the desired result. Wait— we are not done yet! We want to identify the results for year 1970, so we filter the data to the year 1970 and subset the data where the row rank is 2, which is illustrated in step 6. Results are shown in Figure 3-4.

```
+-----+----------+------------+----+
|   id|popularity|release_year|rank|
+-----+----------+------------+----+
|11202|    14.029|        1970|   2|
+-----+----------+------------+----+
```

Figure 3-4. *Output of second highest popularity rating in the year 1970*

We have used the rank function here. There is a subtle difference between rank and denseRank. rank gives the ranking within your ordered partitions, and ties are assigned the same rank. The ranks consecutive to ties are given a higher rank. If you have two rows with rank 2, the next rank will be set at 4. In denseRank, no ranks are skipped and they are consecutive. You have to choose these functions based on your requirements.

Let's take this a step further. What if our question was what is the difference between the revenue of the highest-grossing film of the year and other films within that year? For this we need to calculate the difference in revenue for a film after identifying the highest-grossing film of that year. We will now demonstrate how we can achieve this. Results are shown in Figure 3-5.

```python
# Step 1: Import the window functions

from pyspark.sql.window import *

# Step 2: Select the required subset of columns

df_revenue = df.select('id', 'revenue', 'release_date')

# Step 3: Create the year column from release date

df_revenue=df_revenue.withColumn('release_year',year('release_date')).
drop('release_date')

# Step 4: Define partition function along with range
windowRev =Window.partitionBy(df_revenue['release_year']).orderBy(df_
revenue['revenue'].desc()).rangeBetween(-sys.maxsize, sys.maxsize)

# Step 5: Apply partition function for the revenue differences

revenue_difference = (max(df_revenue['revenue']).over(windowRev) - df_
revenue['revenue'])

# Step 6: Final data

df_revenue.select('id', 'revenue', 'release_year',revenue_difference.
alias("revenue_difference")).show(10,False)
```

```
+-----+---------+------------+------------------+
|id   |revenue  |release_year|revenue_difference|
+-----+---------+------------+------------------+
|665  |1.64E8   |1959        |0.0               |
|10882|5.1E7    |1959        |1.13E8            |
|239  |2.5E7    |1959        |1.39E8            |
|4952 |1.875E7  |1959        |1.4525E8          |
|15944|1.7658E7 |1959        |1.46342E8         |
|213  |1.3275E7 |1959        |1.50724992E8      |
|27029|1.28E7   |1959        |1.512E8           |
|29996|1.22E7   |1959        |1.518E8           |
|11571|1.0E7    |1959        |1.54E8            |
|9660 |6800000.0|1959        |1.572E8           |
+-----+---------+------------+------------------+
only showing top 10 rows
```

Figure 3-5. *Output of revenue differences with the highest revenue by year*

You will observe that there is not much change from the previous example until step 3. In step 4, in addition to the window.partitionBy functions, we have used the rangeBetween function. This is a helpful function to create a boundary between start and end (both inclusive). Both the parameters inside the rangeBetween function indicate to include the lowest and highest possible values within that partition. In step 5, we identify the highest values over the window function and subtract with the row values. This gives the expected output, which is displayed using step 6.

Window functions can be a very long topic. But the preceding examples should give you a fair idea of the possibilities of how you can use the functions. Most SQL windowing functions are available as built-in functions in PySpark.

Other Useful Functions

In this section, we will try to uncover some of other useful functions that we didn't mention in the earlier sections. They can be very effective based on the use cases.

Collect List

We know from previous analysis that some of the titles are repeated. Let's say we want to find all years where the title *The Lost World* is repeated. We have a handy function to do so called collect_list. For the purpose of this demonstration, we will use the existing DataFrame. The previous output is shown in Figure 3-6 for reference. New output is shown in Figure 3-7.

```
+--------------------+-----+
|title               |count|
+--------------------+-----+
|Les Misérables      |8    |
|The Three Musketeers|8    |
|Cinderella          |8    |
|Hamlet              |7    |
|The Island          |7    |
|Dracula             |7    |
|A Christmas Carol    |7    |
|Frankenstein        |7    |
|Framed              |6    |
|The Lost World      |6    |
+--------------------+-----+
only showing top 10 rows
```

Figure 3-6. *Output of filtered and sorted version with one-way frequency*

```
# Step 1: Create the year column from release date

df=df.withColumn('release_year',year('release_date'))

# Step 2: Apply collect_list function to gather all occurrences

df.filter("title=='The Lost World'").groupby('title').agg(collect_
list("release_year")).show(1,False)
```

```
+--------------+------------------------------------------+
|title         |collect_list(release_year)                |
+--------------+------------------------------------------+
|The Lost World|[1999, 2001, 1925, 1960, 1992, 1998]|
+--------------+------------------------------------------+
```

Figure 3-7. *Output of collect_list*

If we break up step 2, we are filtering out the title we are interested in and grouping the DataFrame by the title. This particular title has six occurrences. The collect_list function gathers the release years of this particular title, which is displayed in the output.

Sampling

Sampling holds a special place in analytics. Simply put, a sample is a subset of the population. Okay, what is a population? A population is the entirety of a data. When thinking about a population you are interested in studying, you must be precise. For example, if we say our population is all active users of Amazon, this would imply all users of Amazon irrespective of region, language, or gender. In real-world scenarios, finding population data is difficult and sometimes impractical.

71

We use samples to infer what is happening in the population. If sampling is not done right, it will have a significant impact on your results. If we have to broadly classify the methods available for sampling, you can categorize them into probabilistic and non-probabilistic techniques. Probabilistic use random selection to select the sample. Non-probabilistic techniques are based on subjective judgement. We use probabilistic techniques in data science to make informed decisions on samples. Some of the popular sampling techniques available in probabilistic methods are simple random sampling, systematic random sampling, stratified random sampling, and cluster sampling.

In simple random sampling, every observation in the sample is randomly selected, so all the observations are equally likely to be selected. There are two variants of simple random sampling.

- Simple random sampling with replacement

  In this method, once an observation is randomly selected from the population, it is put back into the dataset before a second observation is selected. In other words, selection of one observation will not affect the outcome of other observations.

- Simple random sampling without replacement

  In this method, once an observation is randomly selected from the population, it is not put back into the dataset before a second observation is selected. In other words, selection of one observation will dramatically affect the outcome of other observations.

In stratified sampling, you have the ability to select the sample from a population that can be partitioned into subpopulations. Say, for example, we know the proportion of the population with certain makes of car in the luxury car segment. When we select the sample, we can enforce the same proportion using this technique.

Let's now see how we can implement these techniques using PySpark.

```
# Simple random sampling in PySpark without replacement

df_sample = df.sample(False, 0.4, 11)
df_sample.count()
```

Output:

```
17711
```

In the preceding command, we are specifying three parameters. The first parameter True/False indicates whether you would like to do a sample with or without replacement. Here, we would like to do it without replacement, so we selected False. The second parameter is the fraction. It indicates the proportion of the population you would like to have in the sample. The third parameter is the seed, which guarantees you the same result when you run this snippet every single time.

```
# Simple random sampling in PySpark with replacement
df_sample = df.sample(True, 0.4, 11)
df_sample.count()
```

Output:

```
17745
```

Note in both cases the sample size is not exactly the ratio that we requested. It is because internally Spark uses a Bernoulli sample technique for taking a sample. Because each element of the population is selected separately for the sample, the sample size is not fixed but rather follows a binomial distribution.

```
# Stratified sampling in PySpark
df_strat = df.sampleBy("release_year", fractions={1959: 0.2, 1960: 0.4,
1961: 0.4}, seed=11)
df_strat.count()
```

Output:

```
260
```

Stratified sample syntax is a bit different from the simple random sample. Observe we are using the sampleBy function instead of sample. The first parameter here is the column you would like to stratify. In our case, we are using the release year. Hypothetically, let's say we know that twice the number of movies were released in 1960 and 1961 when compared to 1959; we would use that fraction to define our sample. This is represented in the second parameter.

Caching and Persisting

Caching is one of the most important concepts of Spark. If used correctly, you can save a lot of time when running large programs. What is *caching*? It is a way to save data temporarily in memory. It is common across different operating systems, software, and programming languages. Why is it useful? If you can save your data in memory, you don't have to repeat the computations to arrive at those results every single time the DataFrame is accessed. You will save the costs of expensive data I/O operations if you use caching effectively. When should you use it? If you are going to access a DataFrame multiple times for multiple operations, we recommend you use caching. Don't use it if you use the DataFrame only once or if data can be computed easily.

After reading your data, you can cache the first query on the DataFrame. The first query will run at a usual speed. But you will experience enhanced computational speed in subsequent operations. Caching is lazy, and that is the reason your first operation on the DataFrame will take time. You can cache any DataFrame using a cache function.

```
df.cache()
```

Persisting has a similar function as or can be seen as a synonym of caching. They differ in the storage-level settings. Spark has five types of storage. This is obtained from the Spark documentation. See Table 3-2.

MEMORY_ONLY

Store RDD as deserialized Java objects in the JVM. If the RDD does not fit in memory, some partitions will not be cached and will be recomputed on the fly each time they're needed. This is the default level.

MEMORY_ONLY_SER

Store RDD as serialized Java objects (one byte array per partition). This is generally more space-efficient than deserialized objects, especially when using a fast serializer, but more CPU-intensive to read.

MEMORY_AND_DISK

Store RDD as deserialized Java objects in the JVM. If the RDD does not fit in memory, store the partitions that don't fit on disk, and read them from there when they're needed.

MEMORY_AND_DISK_SER

Similar to MEMORY_ONLY_SER, but spill partitions that don't fit in memory to disk instead of recomputing them on the fly each time they're needed.

DISK_ONLY

Store the RDD partitions only on disk.

Cache will use `memory_only`. You can use persist for all the levels of storage using the persist function:

```
df.persist()
```

Table 3-2. *Storage Levels in Spark*

Level	Space Used	CPU Time	In Memory	On Disk	Serialized
MEMORY_ONLY	High	Low	Yes	No	No
MEMORY_ONLY_SER	Low	High	Yes	No	Yes
MEMORY_AND_DISK	High	Medium	Partial	Partial	Partial
MEMORY_AND_DISK_SER	Low	High	Partial	Partial	Yes
DISK_ONLY	Low	High	No	Yes	Yes

If you want to remove the data in the storage level manually, you can use the unpersist function:

```
df.unpersist()
```

Saving Data

After all the data operations, you have multiple options to save your dataset, either as hive table or as a text file. Let's first see an example of how we can save the output DataFrame as a text file using `write` and `save`. In the following command, we can specify the format of the output file along with the option of setting a delimiter. We choose a pipe delimited for demonstration, but you can customize and choose any meaningful delimiter that makes sense for your dataset.

```
df.write.format('csv').option('delimiter', '|').save('output_df')
```

Suppose we are working on a cluster and have multiple partitions of the data; we can save them together in a single file using a `coalesce` function. In that case, we would change the preceding command in the following way:

```
df.coalesce(1).write.format('csv').option('delimiter', '|').save('output_
coalesced_df')
```

If we have to partition the data by any column for easy indexing, we can do that while saving using the function `partitionBy`, as shown here:

```
df.write.partitionBy('release_year').format('csv').option('delimiter',
'|').save('output_df_partitioned')
```

If you would like to save your data as a hive table, you can use the `saveAsTable` function:

```
df.write.saveAsTable('film_ratings')
```

Pandas Support

Yes, we are working with PySpark. But there might be times when you want to convert your PySpark DataFrame to a pandas DataFrame. You can do so via the `toPandas()` function. Warning: Do not try to run the following command on large datasets, as you will run into memory issues. This is because you are copying your distributed dataset to local memory. Once a pandas DataFrame is created, there is no connection between the newly created pandas DataFrame and Spark.

```
# Pandas to PySpark

df_pandas=df.toPandas()

# Pandas to PySpark
df_py = spark.createDataFrame(df_pandas)
```

Joins

Figure 3-8 is a quick pictorial representation of all joins. You will learn about each join and how it works in the following section.

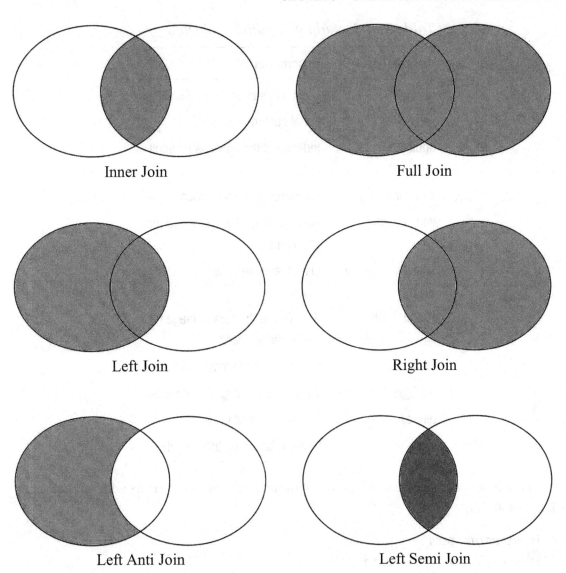

Figure 3-8. *Types of joins*

Joins are by far one of the most frequent operations you do with any DataFrames. For the purpose of joins, let me introduce you to another dataset that has some additional information, which we will work with in conjunction with the data we have been working with so far. See Table 3-3.

Table 3-3. *Metadata of Additional Data*

Variable	Description
id	Unique identifier for the movie
cast	list of cast members
adult	Indicator if the movie is an adult one or not
directors	Film director's information
vote_count	Total number of votes casted for popularity
spoken_languages	List of languages used in the movie
Poster_path	Path where the poster image for the movie is stored
homepage	Web page of the movie if exists
imdb_id	IMDB ID pertaining to the movie
genres	List of genres of the movie
video	Video indicator on the website

Let's first load this data as a new DataFrame and select all the rows with non-missing IDs, as follows:

```
# Input parameters
file_location = "movie_data_part2.csv"
file_type = "csv"
infer_schema = "False"
first_row_is_header = "True"
delimiter = "|"

# Bring all the options together to read the csv file

df_p1 = spark.read.format(file_type) \
.option("inferSchema", infer_schema) \
.option("header", first_row_is_header) \
```

```
.option("sep", delimiter) \
.load(file_location)
```

```
# Cast identifer into the right datatype
```

```
df_p1 = df_p1.withColumn('id',df_p1['id'].cast("integer"))
```

```
# Filter missing values
df_p1=df_p1.filter((df_p1['id'].isNotNull()) & (~isnan(df_p1['id'])))
```

Right away the common variable you will observe in both datasets is ID. We will be able to join both the DataFrames using this identifier. There are multiple joins available in PySpark, so let's go through each of them.

- **Inner join:** This joins both DataFrames on the identifier and retains only the common rows and columns (Figure 3-9).

```
# Inner Join
```

```
df.join(df_p1, df['id'] == df_p1['id'],'inner').printSchema()
df.join(df_p1, df['id'] == df_p1['id']).count()
```

```
root
 |-- id: integer (nullable = true)
 |-- budget: float (nullable = true)
 |-- popularity: float (nullable = true)
 |-- release_date: date (nullable = true)
 |-- revenue: float (nullable = true)
 |-- title: string (nullable = true)
 |-- mean_popularity: double (nullable = false)
 |-- varaiance: double (nullable = true)
 |-- release_year: integer (nullable = true)
 |-- cast: string (nullable = true)
 |-- adult: string (nullable = true)
 |-- directors: string (nullable = true)
 |-- vote_count: string (nullable = true)
 |-- spoken_languages: string (nullable = true)
 |-- poster_path: string (nullable = true)
 |-- homepage: string (nullable = true)
 |-- imdb_id: string (nullable = true)
 |-- genres: string (nullable = true)
 |-- video: string (nullable = true)
 |-- id: integer (nullable = true)
```

```
43783
```

Figure 3-9. *Output columns and counts of inner join*

Observe that Spark has processed the join of the DataFrames and included the identifier twice. As a good practice, you can either rename the identifier with the same name or drop the duplicate identifier altogether.

By default, Spark uses an inner join if no parameter is passed.

- **Left/Left outer join:** A left join merges both the DataFrames on the identifier and retains all the rows of the left-hand DataFrame and matches any rows that are common with the right-hand DataFrame. If there is no equivalent row in the right-hand DataFrame, Spark will insert a null for all the columns.

  ```
  # Left Join
  df.join(df_p1, df['id'] == df_p1['id'],'left').count()
  ```

  Output:

  43998

- **Right/Right outer join:** This merges both DataFrames on the identifier and retains all the rows of the right-hand DataFrame and matches any rows that are common with the left-hand DataFrame. If there is no equivalent row in the left-hand DataFrame, Spark will insert a null for all the columns.

  ```
  # Right Join
  df.join(df_p1, df['id'] == df_p1['id'],'right').count()
  ```

  Output:

  43783

- **Full outer join:** This joins both DataFrames on the identifier and retains all the rows of both the left and right DataFrames that have the same identifier. If there are no equivalent rows in either of the DataFrames, Spark will insert a null for all the columns.

  ```
  # Outer Join
  df.join(df_p1, df['id'] == df_p1['id'],'outer').count()
  ```

  Output:

  43998

- **Left Anti join:** This joins both the DataFrames on the identifier and retains all the rows of left-hand DataFrame that are present in the right-hand DataFrame. It also only retains the left-hand DataFrame schema (Figure 3-10).

```
# Left Anti Join
df.join(df_p1, df['id'] == df_p1['id'],'left_anti').printSchema()
df.join(df_p1, df['id'] == df_p1['id'],'left_anti').count()
```

```
root
 |-- id: integer (nullable = true)
 |-- budget: float (nullable = true)
 |-- popularity: float (nullable = true)
 |-- release_date: date (nullable = true)
 |-- revenue: float (nullable = true)
 |-- title: string (nullable = true)
 |-- mean_popularity: double (nullable = false)
 |-- varaiance: double (nullable = true)
 |-- release_year: integer (nullable = true)

215
```

***Figure 3-10.** Output columns and counts of left anti join*

- ***Left Semi join:*** This is similar to an inner join, except it would not yield the columns from the right-hand DataFrame (Figure 3-11).

```
# Left Semi Join
df.join(df_p1, df['id'] == df_p1['id'],'left_semi').printSchema()
df.join(df_p1, df['id'] == df_p1['id'],'left_semi').count()
```

```
root
 |-- id: integer (nullable = true)
 |-- budget: float (nullable = true)
 |-- popularity: float (nullable = true)
 |-- release_date: date (nullable = true)
 |-- revenue: float (nullable = true)
 |-- title: string (nullable = true)
 |-- mean_popularity: double (nullable = false)
 |-- varaiance: double (nullable = true)
 |-- release_year: integer (nullable = true)

43783
```

***Figure 3-11.** Output columns and counts of left anti join*

- **Broadcasting:** There will be situations where you have to join one large dataset (millions of rows) and one small dataset (hundreds of rows). To join them efficiently, Spark has another function called broadcast. The smaller DataFrame will be replicated in all the nodes of the cluster. This improves efficiency, as Spark doesn't need to do any shuffling in the larger dataset. Here is a working example:

```
# Broadcast Join
```

```
df.join(broadcast(df_p1), df['id'] == df_p1['id'],'left_semi').
count()
```

Output:

43998

Dropping Duplicates

When working with real-world data, another handy and useful function that you may end up using is dropDuplicates. You can drop a duplicate of a single or multiple rows. Say we want to drop all the duplicates in our DataFrames where the title and year are the same; we can do so using the following command:

```
# Dropping Duplicates
df.dropDuplicates(['title','release_year']).count()
```

Output:

43643

```
# Original Data count
```

```
df.count()
```

Output:

43998

Data Visualizations

Well, the sad news is that there are no built-in data visualization packages in Spark. Then what are we discussing here? Okay, we are aware that Python has a rich library of data visualization packages such as matplotlib and seaborn. We can use these packages in concert with the DataFrames and create the required plots.

If we take a step back, we see that we use aggregated data mostly for visualization. If I have to understand a continuous variable, the best plot I can use is a histogram. Similarly, for categorical data I would like to see the group plot. So, we can use the power of Spark to calculate these aggregations and then top it off with the nice visualization packages that are available in Python.

First, let's plot the histogram of popularity data (Figure 3-12):

```
# Step 1: Importing the required libraries

%matplotlib inline
import pandas as pd
from matplotlib import pyplot as plt
import seaborn as sns

# Step 2: Processing the data in Spark. We can use the histogram function
from the RDD
histogram_data = df.select('popularity').rdd.flatMap(lambda x:
x).histogram(25)

# Step 3: Loading the Computed Histogram into a pandas DataFrame for
plotting
hist_df=pd.DataFrame(list(zip(*histogram_data)), columns=['bin',
'frequency'])

# Step 4: Plotting the data
sns.set(rc={"figure.figsize": (12, 8)})
sns.barplot(hist_df['bin'],hist_df['frequency'])
plt.xticks(rotation=45)
plt.show()
```

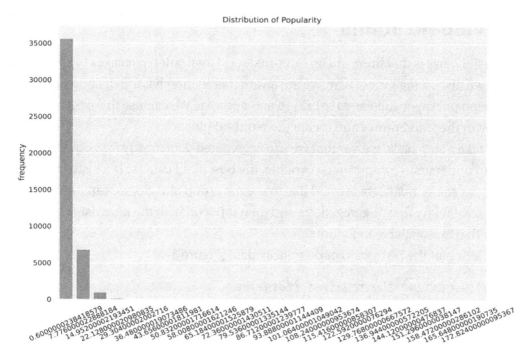

Figure 3-12. *Histogram of popularity*

Let's break down what is happening in each step. First, we imported the required libraries. We are doing all the heavy lifting pertaining to the data in step 2. We are making use of the histogram function available in the RDD API to discretize the distribution and calculate the bins. In this procedure, we requested that it discretize our data into 25 buckets. In step 3, we are creating a pandas DataFrame on top of the data and then using it for plotting. We did a hack here. We are not directly using the histogram functions in Python; rather, all calculations are done in Spark. We are using a bar plot on top of the histogram data to look at the distribution. As you can observe, the distribution is hard to interpret because of the sparseness of the values in the tail.

Let's try filtering the data to see if we can get a better idea of the distribution.

```
# Filtering the data to get a better understanding of data
df_fil=df.filter('popularity<22')
```

```
# Processing the data in Spark. We can use the histogram function from the
RDD
histogram_data = df_fil.select('popularity').rdd.flatMap(lambda x:
x).histogram(25)
```

```
# Loading the computed histogram into a pandas DataFrame for plotting
hist_df=pd.DataFrame(list(zip(*histogram_data)), columns=['bin',
'frequency'])
```

```
# Plotting the data
sns.set(rc={'figure.figsize':(11.7,8.27)})
sns.barplot(hist_df['bin'],hist_df['frequency'])
plt.xticks(rotation=25)
plt.title('Distribution of Popularity - Data is filtered')
plt.show()
```

We repeated the same steps except for filtering. Now we see a much better
distribution (Figure 3-13).

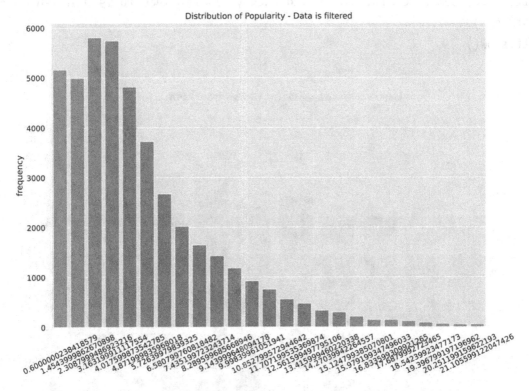

Figure 3-13. *Histogram of popularity after filtering*

Let's try to plot another variable. Say we want to know how many films were released
between 1960 and 1970 by the year in our dataset. We can do it using the following
commands (Figure 3-14):

85

```
# Step 1: Preparing the data using Spark functions and converting to pandas
DataFrame

df_cat=df.filter("(release_year>1959) and (release_year<1971)").
groupby('release_year').count().toPandas()

# Step 2: Sorting the values for display

df_cat=df_cat.sort_values(by=['release_year'], ascending=False)

# Step 3: Plotting the data
sns.set(rc={'figure.figsize':(11.7,8.27)})
sns.barplot(df_cat['release_year'],df_cat['count'])
plt.xticks(rotation=25)
plt.title('Number of films released each year from 1960 to 1970 in our
dataset')
plt.show()
```

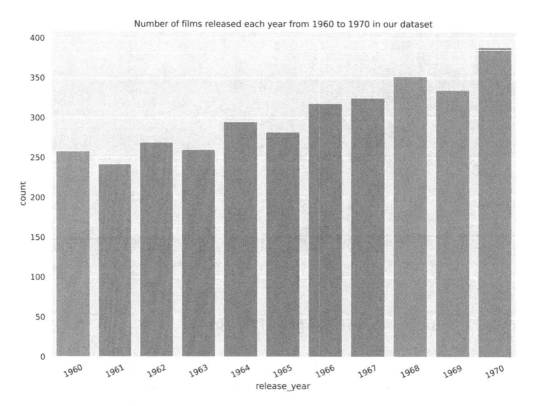

Figure 3-14. *Number of films released every year from 1960 to 1970*

Use the following dataset for Exercise 3-1: `https://www.kaggle.com/kakarlaramcharan/tmdb-data-0920`.

EXERCISE 3-1: DATA MANIPULATIONS

Question 1: Identify the second most popular movie in 2010 based on popularity.

Question 2: Identify all title names that are repeated and show the years in which they were repeated.

Question 3: Identify the top movies by popularity across all years.

Question 4: Visualize the budget of movies in 2014 and 2015.

Introduction to Machine Learning

Before we jump into the world of machine learning, let's understand the progression of analytics. Analytics is a computational analysis of data or statistics. It is used for the discovery, interpretation, and communication of meaningful patterns of data. You might hear multiple terms and buzz words related to analytics, but at the end of the day they are all used for collecting insights and making informed decisions.

Analytics broadly can be divided into the following (Figure 3-15):

> **Descriptive:** This field of analytics is focused on creating reports and charts understanding the data.

> **Predictive:** This field of analytics is associated with estimating what will happen given what has happened. It relies on historical data to make predictions.

> **Prescriptive:** This field of analytics is focused on what has to be done given the past data. This includes decision optimization.

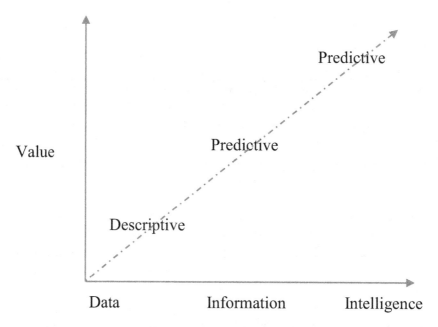

Figure 3-15. *Overview of analytics*

Often times these terms are used in a way that can be confusing to new learners. Sure, there is an overlap. Machine learning is a subfield of computer science and is considered a subset of artificial intelligence. *Predictive analytics* refers to statistical modeling and is a term if you commonly find in the insurance, banking, and marketing industries. *Machine learning* is a broader term that encompasses multiple categories, including the following:

> **Supervised learning:** In this type of learning, an algorithm is provided a historical dataset with target labels or events. This field of study is validated through measures of accuracy. This also includes any time-series modeling.

> **Unsupervised learning:** In this type of learning, an algorithm is provided a historical dataset with no target labels or events. The intent here is to identify the hidden patterns or segments in the data.

Let's understand the data types before we discuss more about the types of learning techniques found in the supervised and unsupervised methods. We can broadly classify data into qualitative and quantitative.

Qualitative data is something that can't be measured. This can include color of your vehicle, marital status, education, etc. These variables can also be called categorical variables. We can further classify the qualitative data into the following:

- Nominal – When there are multiple levels in the data and no natural ordering, you can consider it to be nominal. E.g., color of your vehicle, marital status

- Ordinal – When there is an inherent ordering in your data, you can consider it to be ordinal. E.g., Education. In education you can have multiple levels, such as less than high school, high school, graduate, and postgraduate. These levels have a ranking associated with them. We know postgraduate is on the higher end of the spectrum, while less than high school is on the lower end of the spectrum.

Quantitative data can be measured. This can include age, distance, etc. Quantitative data can be further divided into discrete and continuous.

- Discrete – You can consider data to be discrete when it has integer values; e.g., age (10, 20)

- Continuous – You can consider data to be continuous when it has fractional or decimal data; e.g., distance (40.71 miles, 59.82 miles)

We have dedicated a chapter to unsupervised learning/clustering in which you will learn different techniques and methods (Figure 3-16).

Why is it important to know these datatypes? Well, because it will help you determine what types of algorithms can be applied. Certain algorithms are designed for selected datatypes. In supervised learning, there are two different classes of algorithms: classification and regression.

Classification: In this class of machine learning, the target/event data is discrete. Examples include predicting the customers who are likely to churn (0/1) or the customers who are likely to default on a loan (0/1).

Regression: In this type of machine learning, the target/event data is continuous. Examples include predicting the price of a house ($) based on given parameters or predicting the purchases of customers in a given month.

We will explore in detail in later chapters which algorithms can be applied to each of these methods.

Clustering can be defined as grouping homogeneous groups of data together in a dataset.

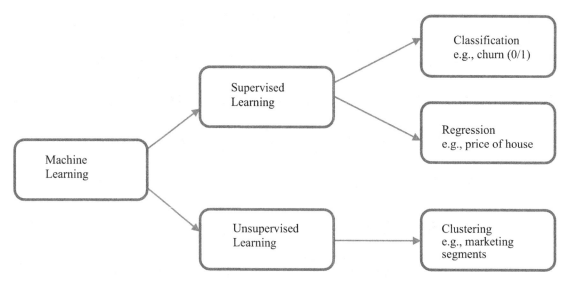

Figure 3-16. *Classification of machine learning*

There are standard methodologies most of data professionals follow for predictive modeling. It is called CRISP–DM, or Cross-Industry Standard for Data Mining (Figure 3-17). It consists of six major milestones, as follows:

1. **Business Understanding:** This is the first important step to understanding the problem statement and formulating it into a data problem. This comes from the business.

2. **Data Understanding:** This step includes data collection and descriptive data analysis to discover the data and get familiar with interesting features.

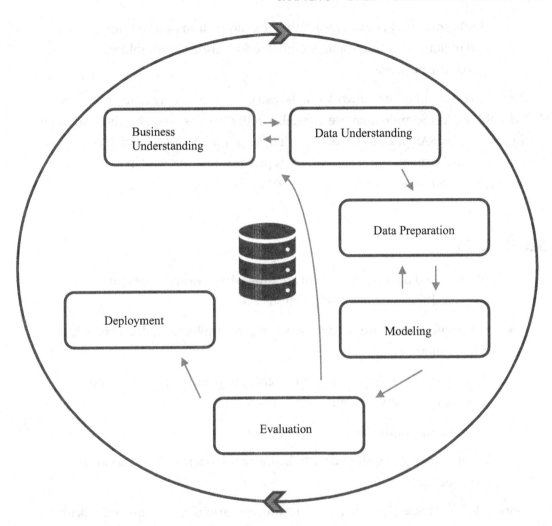

Figure 3-17. *CRISP–DM Methodology*

3. **Data Preparation:** This includes consolidating all the data features and creating additional features/variables from the dataset.

4. **Modeling:** This includes applying appropriate algorithms and looping to create more variables if necessary.

5. **Evaluation:** Once one or multiple models are built, a model is selected based on validation criteria set. We have a separate chapter that will help you understand all the model evaluation and validation techniques. We also seek feedback from the business as to whether the model is doing the intended function.

6. **Deployment:** This involves putting the model into production and making the model scores available for various stakeholders across the business.

An alternate standard framework that is also common among data practitioners is SEMMA. SEMMA is Sample, Explore, Model, Modify, and Assess. This methodology was introduced by SAS prior to CRISP–DM. This is a modeling-focused approach and does not include business inputs as part of its process. It works well in decentralized organizations if you have a clear definition of your business objective.

Summary

- We explored the possibilities of creating window functions and answered specific business questions.

- We learned what sampling is, how it can be implemented, and why it is important.

- We learned the mechanics of the `cache` and `persist` functions and when they should be used.

- We learned simple hacks for visualizations.

- We also are now equipped with the basics of machine learning and its framework.

Good job! In the next chapter, we will introduce variable selection and talk about different techniques. When you are inundated with data, you can always count on the variable selection chapter. It will help you identify the top variables with information. See you in the next chapter!

CHAPTER 4

Variable Selection

Variable selection is an art. The goal of this chapter is to help you understand the different variable selection techniques that can be used to select the best features in your dataset. It is one of the key processes in data science. To put it in simple terms, let us say you are the coach for a soccer team. You want to pick the best team to win the World Cup. You need to have the best player in each position (*best features*), and you don't want too many players who play the same position (*multicollinearity*) at a given point in time. In PySpark, there are multiple ways to accomplish this task. In this chapter, we will go through some built-in and custom-built variable selection techniques (Figure 4-1).

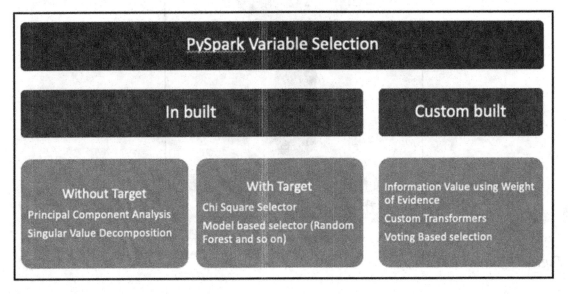

Figure 4-1. *Variable selection*

© Ramcharan Kakarla, Sundar Krishnan and Sridhar Alla 2021
R. Kakarla et al., *Applied Data Science Using PySpark*, https://doi.org/10.1007/978-1-4842-6500-0_4

In the upcoming pages, we will explore the concepts behind each technique, do some calculations by hand to see how they work *(whenever necessary)*, and finally implement them using PySpark. For this entire chapter, we will use the bank dataset from the UCI machine learning repository at `https://archive.ics.uci.edu/ml/ datasets/Bank+Marketing`. You need to download the *bank.zip* file from the *data* folder to access the *bank-full.csv* file. Before we proceed further, here is an intuitive way to look at variable selection (Figure 4-2).

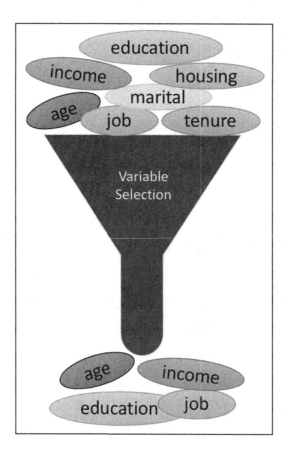

Figure 4-2. *Intuitive way*

Exploratory Data Analysis

Exploratory data analysis (EDA) is a process used to analyze data so as to identify the characteristics and patterns existing in that data using summarization or visualization. In the previous chapters, we have gone through a lot of code to cover this in detail. We recommend you use these pieces of code on any dataset that you work with. When you understand the data better, you can make quick decisions and explain the reasoning behind your decisions too. EDA can also become a powerful first step in the variable selection process. If you wonder how, think about cardinality (see definition in this section) and missing values. It is good practice to check these two things when doing any machine learning exercise.

Let's use this example of home prices to better understand cardinality and missing values (Figure 4-3).

Zip	Home type	HOA fees	Sq. Ft	Price
19102	Town Home	$100	2400	$350K
19103	Condo	$300	1200	$200K
19145	Single Family Unit	NaN	2500	$400K
19150	Single Family Unit	NaN	1800	NaN
19040	Town Home	$250	2275	$325K

Figure 4-3. *Home prices dataset*

Cardinality

Cardinality is the number of unique values of a variable. Let's assume that we are building a model to predict house prices, and let's assume that the home type is "Single Family Unit" for all five records. In this case, the cardinality for the *Home type* variable is 1 ("Single Family Unit"). In this case, *Home type* is not contributing any useful information, because, any record you choose, this variable is going to give you the same information ("Single Family Unit"). Therefore, we can omit this variable and use other variables to build the predictive model. Although you can calculate the cardinality of any variable, cardinality is useful for categorical variables. In Figure 4-3, the cardinality of *Home type* is 3.

Missing Values

Missing values, as the name suggests, refers to a missing piece of information. We need to understand the reason behind the missing piece of information in order to build a better model. In general, data could be missing for the following three reasons:

- Missing at random (MAR)

- Missing completely at random (MCAR)

- Missing not at random (MNAR)

By understanding the type of missing information, we can treat the data accordingly. We will use the home price example to go through in detail.

Missing at Random (MAR)

When a piece of data is missing at random, there is a relationship between the missing data and the observed data. Let us use our home price example to understand this better. When you look closely at the *Home type* and *HOA fees* (Homeowners Association fees) variables, you can see that there is a value for *HOA fees* when *Home type* is "Town Home" or "Condo" and the fee is missing when the *Home type* is "Single Family Unit." This makes sense because a single family unit (in most cases) does not have any HOA fees. In this case, *HOA fees* for "Single Family Unit" homes can be imputed with zero. This type of missing data is called *missing at random (MAR)*.

Missing Completely at Random (MCAR)

When a piece of data is missing completely at random, there is no relationship between the missing data and the observed data. Let us look at the *Price* variable in our data. One of the home prices is missing. Unlike the MAR example we just discussed, there is no logic that we can derive for this missing data. When you start to think about the reason behind the missing information, there could be a lot. Maybe the house is listed for auction or the house is a new construction and the seller is yet to determine a price, or there could be some other reason. In any case, the missing price has nothing to do with any of our observed data. This is a typical example for data *missing completely at random (MCAR)*. In order to impute this data, you might have to use your own logic. Imputing is a way of filling in missing values in data science literature. We do have some recommendations for how to treat this type of missing data, as follows:

- Mean, median, or mode imputation

- Model-based imputation (this will be discussed in future chapters)

- Multiple imputation

- Use business logic to impute the missing information

- Drop features with significant missing data (variable selection)

- Drop rows with missing values (least recommended)

How would one use business logic to impute the missing information? Let's assume you work for Zillow and you know that the *price* information is missing only if the house is listed for auction. In this case, your team recommends you use a default price for houses listed in auction (let's say $1). Then, you would impute the price with $1.

Missing Not at Random (MNAR)

When a data is missing not at random, it means that the data is not collected. In our example, we do not have information on when the house is built (*year*). This information is missing because we did not account for the *year* variable in our data collection process. To resolve this issue, you should understand and improve the data collection process to account for the missing information.

We have discussed the types of missing values in detail. Can we use this to reject variables in our dataset? Yes, we can. Most of the data science folks reject a variable based on the percentage of the data missing. For example, you can reject a variable if it has more than 80 percent missing values. The 80 percent is not a standard value, and it could vary on a case-to-case basis.

EXERCISE 4-1: MISSING VALUES & CARDINALITY

Download the Melbourne Housing dataset: `https://www.kaggle.com/dansbecker/melbourne-housing-snapshot/data#`.

Question 1: Calculate the cardinality for each variable in the dataset. Can you reject any variables based on cardinality? (Hint: Look for cardinality of 1)

Question 2: Calculate the missing value percentage for each variable in the dataset. Can you eliminate any variables based on missing values? (Hint: Just for this exercise, reject any variable that has more than 45 percent missing values)

Question 3: Impute the *YearBuilt* column with median value. (Hint: Use 0.1 for `relativeError`)

Question 4: Impute *BuildingArea* column to create a new variable called `mean_imputed_BuildingArea` and calculate the new mean after imputation.

Coming back to the bank dataset, let us load the dataset and do some EDA-based variable selection.

```
filename = "bank-full.csv"
target_variable_name = "y" #This variable can be set later too
```

Let's load the data using the following code:

```
from pyspark.sql import SparkSession
spark = SparkSession.builder.getOrCreate()
df = spark.read.csv(filename, header=True, inferSchema=True, sep=';')
df.show()
```

You need to make sure to provide a semicolon (;) as the separator for this dataset, or it won't work. In addition, we see a new option inferSchema in the file load operation. When this option is enabled, it automatically infers the datatype of each column in the dataset. Figure 4-4 shows the schema inference with and without this option.

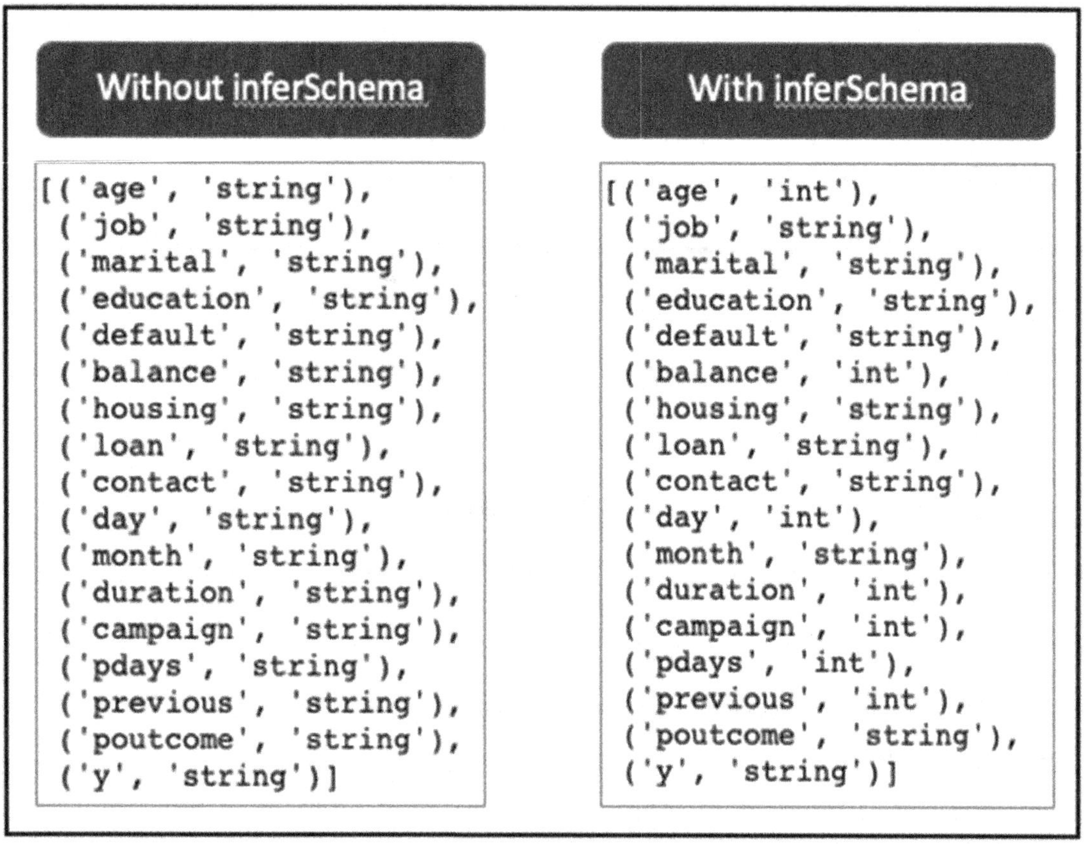

Figure 4-4. *Infer schema option output difference*

Okay, we have the data loaded in our environment. Here are some codes to get you started to play with the data

```
#Length of the data
df.count() #45211

#describes data
df.describe().toPandas() #missing and cardinality followup

#type of each variable
df.dtypes
df.printSchema()

#single variable count by group
df.groupBy('education').count().show()

#target count
df.groupBy(target_variable_name).count().show()

#multiple column group by
df.groupBy(['education',target_variable_name]).count().show()

#column aggregations
from pyspark.sql.functions import *
df.groupBy(target_variable_name).agg({'balance':'avg', 'age': 'avg'}).
show()
```

	summary	age	job	marital	education	default
0	count	45211	45211	45211	45211	45211
1	mean	40.93621021432837	None	None	None	None
2	stddev	10.618762040975401	None	None	None	None
3	min	18	admin.	divorced	primary	no
4	max	95	unknown	single	unknown	yes

Figure 4-5. *Column aggregation summary*

We want to highlight the described data output shown in Figure 4-5 and connect it to the cardinality and missing values that we discussed earlier. The *count* row in the *summary* column gives the count of non-missing values in a variable. The difference between the total observations and this column gives you the total number of missing values in each column. Let us go over two more stats here: the *min* and *max* rows in the *summary* column. When the *min* and *max* rows have the same value for a variable, then the cardinality of the variable is 1. If you are unable to connect, no worries. You can use the following codes to perform these checks.

Code 1: Cardinality Check

```python
# Cardinality Check
from pyspark.sql.functions import approxCountDistinct, countDistinct
"""
Note: approxCountDistinct and countDistinct can be used interchangeably.
Only difference is the computation time.
"approxCountDistinct" is useful for large datasets
"countDistinct" for small and medium datasets.
"""
def cardinality_calculation(df, cut_off=1):
    cardinality = df.select(*[approxCountDistinct(c).alias(c) for c in
    df.columns])

    ## convert to pandas for efficient calculations
    final_cardinality_df = cardinality.toPandas().transpose()
    final_cardinality_df.reset_index(inplace=True)
    final_cardinality_df.rename(columns={0:'Cardinality'}, inplace=True)

    #select variables with cardinality of 1
    vars_selected = final_cardinality_df['index']
    [final_cardinality_df['Cardinality'] <= cut_off]

    return final_cardinality_df, vars_selected

cardinality_df, cardinality_vars_selected = cardinality_calculation(df)
```

Code 2: Missing Values Check

```
#missing values check
from pyspark.sql.functions import count, when, isnan, col

# miss_percentage is set to 80% as discussed in the book
def missing_calculation(df, miss_percentage=0.80):
    #checks for both NaN and null values
    missing = df.select(*[count(when(isnan(c) | col(c).isNull(), c)).
            alias(c) for c in df.columns])
    length_df = df.count()
    ## convert to pandas for efficient calculations
    final_missing_df = missing.toPandas().transpose()
    final_missing_df.reset_index(inplace=True)
    final_missing_df.rename(columns={0:'missing_count'}, inplace=True)
    final_missing_df['missing_percentage'] = final_missing_df['missing_
                                           count']/length_df

    #select variables with cardinality of 1
    vars_selected = final_missing_df['index'][final_missing_df['missing_
    percentage'] >= miss_percentage]

    return final_missing_df, vars_selected

missing_df, missing_vars_selected = missing_calculation(df)
```

For the bank dataset, no columns are rejected based on these two checks. However, in real-world datasets, you will see a handful of columns that can be eliminated after performing cardinality and missing value checks. Well, that is all we need to learn for EDA-based variable selection techniques. Before we move ahead with other selection techniques, we will go through one more feature engineering exercise in EDA. This will ensure that our data is ready for downstream tasks. Remember Figure 4-4, where we still have some columns that are strings. Well, the next set of selection techniques requires all the columns to be numbers.

Converting strings to numbers can be achieved using one of the following options:

1. Individually casting columns to desired data type. This is cumbersome for large datasets.

2. OneHotEncoder (for Spark >= 3.0) or OneHotEncoderEstimator (for Spark >= 2.3) from pyspark.ml.feature. This option will create dummy variables for each category, thereby increasing the number of features we work with. In our dataset, the *education* variable has four categories: primary, secondary, tertiary, and unknown. After using OneHotEncoder on this column, the output looks like Figure 4-6. *Note: PySpark by default does not produce output like that shown; it requires some tweaking using* udf *functions discussed in the previous chapter.*

Education	Education_Primary	Education_Secondary	Education_Tertiary	Education_Unkown
primary	1	0	0	0
secondary	0	1	0	0
tertiary	0	0	1	0
Unknown	0	0	0	1
Primary	1	0	0	0
secondary	0	1	0	0

Figure 4-6. *One hot encoder*

3. StringIndexer from pyspark.ml.feature. This option assigns a number to each level. The most frequent value would get index 0, followed by the next most frequent and so on. The output for the *education* variable using StringIndexer would look like Figure 4-7. Although the output of StringIndexer looks like an ordinal output (increasing order), it should not be considered as ordinal. Most algorithms fail to treat this variable as nominal output (mutually exclusive levels). This is one of the drawbacks of StringIndexer.

Education	Education_Index
primary	0
secondary	1
tertiary	2
Unknown	3
Primary	0
secondary	1

Figure 4-7. *StringIndexer output*

4. The last option is to use a weighted index, something like a Weight of Evidence, which will be discussed later in this chapter. It overcomes both the drawbacks faced by StringIndexer and OneHotEncoder. It uses the distribution of target versus non-target population to create the weights for each category.

For our example, let us use StringIndexer to convert all our columns. The code is provided next.

Step 1: Identify Variable Types

```
# Identify variable types
def variable_type(df):
    # use the dtypes to separate character and numeric variables
    vars_list = df.dtypes
    char_vars = [] #character variables list
    num_vars = [] #numeric variable list
    for i in vars_list:
        if i[1] in ('string'):
            char_vars.append(i[0])
        else:
            num_vars.append(i[0])

    return char_vars, num_vars

#apply variable_type function on our DataFrame
char_vars, num_vars = variable_type(df)
```

Step 2: Apply StringIndexer to Character Columns

To the character variables identified in step 1 (char_vars), apply the StringIndexer function. The code is provided here:

```
from pyspark.ml.feature import StringIndexer
from pyspark.ml import Pipeline

#converts each category column to index
def category_to_index(df, char_vars):

    char_df = df.select(char_vars)
    indexers = [StringIndexer(inputCol=c, outputCol=c+"_index",
            handleInvalid="keep") for c in char_df.columns]
    pipeline = Pipeline(stages=indexers)
    char_labels = pipeline.fit(char_df)
    df = char_labels.transform(df)
    return df, char_labels

#apply category_to_index function on our DataFrame
df, char_labels = category_to_index(df, char_vars)
```

A couple of things are highlighted in the code here. The first is the StringIndexer operation, which takes in an input column (*Education*) and produces an output column (*Education_index*). One thing to note is the handleInvalid option, which is set to *keep*, so that when we encounter new data in the future it will still work. The next option is the Pipeline option, which is used to execute our steps sequentially. We will cover pipelines in detail in later chapters. The DataFrame output currently has the old string columns and the new numeric columns. Let's subset our data to the required columns, as follows:

```
df = df.select([c for c in df.columns if c not in char_vars])
```

Finally, we have a DataFrame that is now completely numeric. Folks who are interested in renaming the *_index* variables to the original name can use the following code to perform the operation:

```
#rename _index columns to original variable name
def rename_columns(df, char_vars):
    mapping = dict(zip([i + '_index' for i in char_vars], char_vars))
    df = df.select([col(c).alias(mapping.get(c, c)) for c in df.columns])
```

```
    return df
```

```
# apply rename_columns to our DataFrame
df = rename_columns(df, char_vars)
```

Step 3: Assemble Features

The last step is to assemble the individual variables into a single feature vector. This is useful because, instead of providing individual variables in the next steps, we can point to one variable. In addition, you can optionally scale the DataFrame using a StandardScaler or MinMaxScaler. This is accomplished using the following code:

```
#assemble features into one vector
from pyspark.ml.feature import VectorAssembler
```

```
#assemble individual columns to one column - 'features'
def assemble_vectors(df, features_list, target_variable_name):
    stages = []
    #assemble vectors
    assembler = VectorAssembler(inputCols=features_list,
    outputCol='features')
    stages = [assembler]
    #select all the columns + target + newly created 'features' column
    selectedCols = [target_variable_name, 'features'] + features_list
    #use pipeline to process sequentially
    pipeline = Pipeline(stages=stages)
    #assembler model
    assembleModel = pipeline.fit(df)
    #apply assembler model on data
    df = assembleModel.transform(df).select(selectedCols)
    return df
```

```
# exclude target variable and select all other feature vectors
features_list = df.columns
features_list.remove(target_variable_name)
```

```
# apply the function on our DataFrame
df = assemble_vectors(df, features_list, target_variable_name)
```

If you prefer to look at the schema of the combined vector features, you can use the following code (Figure 4-8):

```
{'numeric': [{'idx': 0, 'name': 'age'},
    {'idx': 1, 'name': 'balance'},
    {'idx': 2, 'name': 'day'},
    {'idx': 3, 'name': 'duration'},
    {'idx': 4, 'name': 'campaign'},
    {'idx': 5, 'name': 'pdays'},
    {'idx': 6, 'name': 'previous'},
    {'idx': 7, 'name': 'job'},
    {'idx': 8, 'name': 'marital'},
    {'idx': 9, 'name': 'education'},
    {'idx': 10, 'name': 'default'},
    {'idx': 11, 'name': 'housing'},
    {'idx': 12, 'name': 'loan'},
    {'idx': 13, 'name': 'contact'},
    {'idx': 14, 'name': 'month'},
    {'idx': 15, 'name': 'poutcome'}]}
```

Figure 4-8. *Features column schema*

```
df.schema["features"].metadata["ml_attr"]["attrs"]
```

To make them as a nice pandas DataFrame, you can use the following code:

```
import pandas as pd
for k, v in df.schema["features"].metadata["ml_attr"]["attrs"].items():
    features_df = pd.DataFrame(v)
```

That's it. We now have a DataFrame (df) that is compatible with the downstream tasks. One thing to notice is the target variable *y* looks like Figure 4-8 after numeric conversion (Figure 4-9). This is due to the fact that the *no* option had more frequency than the *yes* option. You need to remember this so that the results interpretation makes sense. You can also explicitly convert the target to the desired numerical mapping based on personal choice.

```
+---+-----+
|  y|count|
+---+-----+
|0.0|39922|
|1.0| 5289|
+---+-----+
```

Figure 4-9. *Final target mapping*

Quick Exercise   Try using `StandardScaler` with `VectorAssembler` to create a scaled DataFrame in the previous step. (Hint: You need to add the scaler model in the stages, along with assembler)

Built-in Variable Selection Process: Without Target

In a machine learning model setting, you can reduce the dimension of the data without an explicit target. This type of dimensionality reduction is called data compression. Think of an image file. You have multiple ways to compress the file, thereby reducing the file size. In this section, we will discuss two implementations that are readily available in Spark ML to perform these task: PCA and SVD.

Principal Component Analysis

Principal component analysis (PCA) can be used to identify patterns in large datasets. It can recognize the variables in your data that contain the most information. It is a linear representation of your dataset. Today, we have deep learning algorithms like Autoencoder that do the same things and are useful for modeling complex non-linear relationships. For the scope of this book, we will only focus on PCA. A general rule of thumb: If the variable distribution is all centered close to the mean, it usually contains tiny information. Let us understand the PCA mechanics with a simple example.

Mechanics

Let's say we have six customers who have bought the same car with the same model and same make in different months of a year. We want to understand the top variable that can explain their satisfaction (see Table 4-1).

Table 4-1. *Sample Data*

Participant_ID	Quality	Reliability
1	10	6
2	9	4
3	8	5
4	3	3
5	2	2
6	1	1

When we use a single variable *Quality* and plot the data, we observe participants one to three will be close to each other on the upper end of the line, and the remaining three participants will be on the lower end of the line. Even with a single data line we are able to make an inference that the first three participants are similar and the last three participants are similar. When we plot this data in a two-dimensional space, we can see the first three participants cluster to the top right and the last three participants cluster to the bottom left (Figure 4-10a). PCA will guide us in identifying the variable that is most important to the clustering of these variables. Let's embark on the journey of how PCA is calculated.

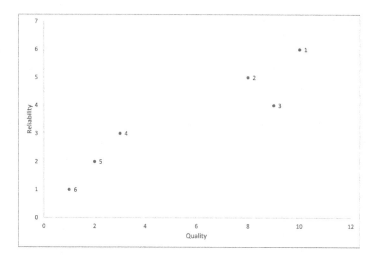

Figure 4-10a. *Car satisfaction data*

First, the means of both the variables are calculated. Mean values of *Quality* and *Reliability* are (5.5,3.5) respectively. We shift the origin from (0,0) as shown in the graph in Figure 4-10b to (5.5,3.5). This is to make sure the data is centered on the origin.

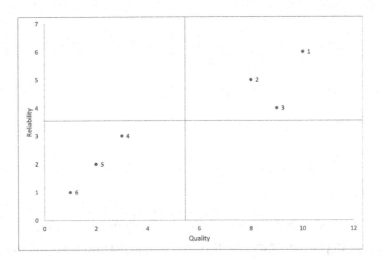

Figure 4-10b. *Car satisfaction data with axis*

Now we try to fit a line through these points until it fits the data well, passing through the origin. How can you define this criterion and know a line is fitting closely? Well, you do know that when the projected distances from the data to the line are minimized, it is the closest fit (Point 5 is projected to the line passing through the origin). The same condition can be flipped and can be stated as a line that maximizes the distance from projected points to the origin (dark shaded line in the same figure). PCA finds the best-fitting line by maximizing the sum of the squared distances from the projected points to the origin (Figure 4-10c).

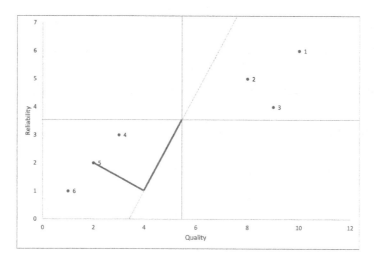

Figure 4-10c. *PCA projections*

The reason we use squared distances is to avoid any negative values. We start the process by rotating an arbitrary line through the origin and start calculating the distances from the origin to the projected points on the line. In the preceding figure, we show the distance of projected Point 5 to the origin. The highlighted dark line indicates the distance. Similar distances are calculated for the remaining points to arrive at the sum of the squared distances. All these distances are compared, and the line that has the *maximum sum of squared distance* is selected (Figure 4-10d).

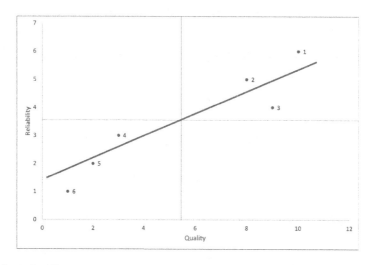

Figure 4-10d. *PCA line*

The indicated line maximizes the sum of the squared distance across all the points. So, this is considered as PC1 or principal component 1. Assuming this line has a slope of 0.25, we can infer that for every four units we move in the quality direction, we will go up in reliability by one unit. This is a pseudo-indication that the data is all spread out on the quality axis. Therefore, quality plays a major role in defining PC1. This component can be thought of as a linear combination of four units of quality and one unit of reliability. PC2 is simply the line through the origin that is perpendicular to PC1. Its ratio of variables would be inverse to that of PC1. This component can be thought of as a linear combination of four units of reliability and one unit of quality. If we scale both of these to unit vectors, we can calculate the *loading scores for PCs. Eigenvalues* are the sum of squares of the distances between the projected points and the origins of the identified principal components. We can covert the *eigenvalues* of each principal component into variations by dividing them by (`sample size -1`).

Here is the code for the PySpark implementation of PCA:

```
from pyspark.ml.feature import PCA
from pyspark.ml.linalg import Vectors
no_of_components = 3 #custom number
pca = PCA(k=no_of_components, inputCol="features", outputCol="pcaFeatures")
model = pca.fit(df)
result = model.transform(df).select("pcaFeatures")
result.show(truncate=False)
```

`pcaFeatures` is the reduced set of features produced by the PCA algorithm that can be used in the downstream machine learning process. Notice, we ran with sixteen input variables, and we ended up with three components. Essentially, we have reduced our dimension size by a factor of 1/5 (3/16). To get the loading scores for each variable, we can use the following code:

```
model.pc.toArray()
array([[-3.41021399e-04,  2.79524640e-04,  2.58353293e-03],
       [-9.99998245e-01,  1.83654726e-03,  1.13892524e-04],
       [-1.22934480e-05,  9.79995613e-04,  7.79347982e-03],
       [-1.83671689e-03, -9.99996986e-01, -7.36955549e-04],
       [ 1.48468991e-05,  1.01391994e-03,  2.75121381e-03],
       [-1.13085547e-04,  7.49207153e-04, -9.99889046e-01],
       [-1.26153895e-05, -6.36100089e-06, -1.04654388e-02],
       [-1.78789640e-05, -4.11817349e-05,  5.51411389e-04],
```

```
[  6.41085932e-06, -5.23364803e-05, -1.45349520e-04],
[-1.11185424e-05,  1.30366514e-05,  2.01500982e-04],
[  2.91665702e-06,  4.42643869e-06,  3.95562163e-05],
[-1.12221341e-05,  1.26153926e-05,  6.17569266e-04],
[  1.01623400e-05,  1.50687571e-05,  8.23933054e-05],
[-5.68377754e-07,  6.95393403e-05,  1.03951369e-03],
[-7.60886236e-05, -1.16754927e-04, -3.24662847e-03],
[-8.55162111e-06, -6.01853226e-05, -4.94522998e-03]])
```

Now, let's map these loading scores to individual variables. We are using the *features_list* variable, which we fed earlier to the `assembler` function.

Table 4-2.

Variable	PC1	PC2	PC3
age	-0.0003	0.0003	0.0026
balance	-1.0000	0.0018	0.0001
day	0.0000	0.0010	0.0078
duration	-0.0018	-1.0000	-0.0007
campaign	0.0000	0.0010	0.0028
pdays	-0.0001	0.0007	-0.9999
previous	0.0000	0.0000	-0.0105
job	0.0000	0.0000	0.0006
marital	0.0000	-0.0001	-0.0001
education	0.0000	0.0000	0.0002
default	0.0000	0.0000	0.0000
housing	0.0000	0.0000	0.0006
loan	0.0000	0.0000	0.0001
contact	0.0000	0.0001	0.0010
month	-0.0001	-0.0001	-0.0032
poutcome	0.0000	-0.0001	-0.0049

You can see that the *balance* variable dominates the PC1 component, *duration* the second PC2 component, and *pdays* the third PC3 component. You should be cautious about these results because the input data is not standardized. The continuous features are dominating the output of PCA since they have high variance. Another output of the algorithm you should focus on is the variance explained by each component. In PySpark, you can use the following code to do that:

```
model.explainedVariance
#Output: DenseVector([0.9918, 0.0071, 0.0011])
```

You can use the following code to plot these values as a graph. This graph is called a *Scree plot.*

```
import matplotlib.pyplot as plt
import numpy as np
x = []
for i in range(0, len(model.explainedVariance)):
    x.append('PC' + str(i + 1))
y = np.array(model.explainedVariance)
z = np.cumsum(model.explainedVariance)
plt.xlabel('Principal Components')
plt.ylabel('Variance Explained')
plt.bar(x, y)
plt.plot(x, z)
```

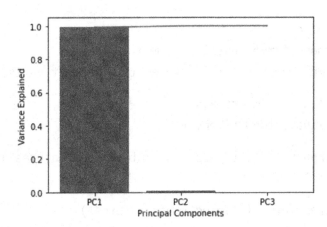

Figure 4-10e. *Variance explained*

As you can see in Figure 4-10e, the first principal component explains 99 percent of the variance in the data. This is because we did not scale our DataFrame before we used PCA. So, let us do one more run with StandardScaler before using PCA. You will notice a big change in the variance explained plot.

```python
from pyspark.ml.feature import VectorAssembler
from pyspark.ml.feature import StandardScaler

#assemble and scale individual columns to one column - 'features2'
def scaled_assemble_vectors(df, features_list, target_variable_name):
    stages = []
    #assemble vectors
    assembler = VectorAssembler(inputCols=features_list,
    outputCol='assembled_features')
    scaler = StandardScaler(inputCol=assembler.getOutputCol(),
    outputCol='features2')
    stages = [assembler, scaler]
    #select all the columns + target + newly created 'features' column
    selectedCols = [target_variable_name, 'features2'] + features_list
    #use pipeline to process sequentially
    pipeline = Pipeline(stages=stages)
    #assembler model
    scaleAssembleModel = pipeline.fit(df)
    #apply assembler model on data
    df = scaleAssembleModel.transform(df).select(selectedCols)
    return df
features_list = df.columns
features_list.remove(target_variable_name)
df = scaled_assemble_vectors(df, features_list, target_variable_name)

from pyspark.ml.feature import PCA
from pyspark.ml.linalg import Vectors

pca = PCA(k=3, inputCol="features2", outputCol="pcaFeatures")
model = pca.fit(df)

result = model.transform(df).select("pcaFeatures")
result.show(truncate=False)
```

When you run PCA with the newly created DataFrame from the preceding code, you see the variance explained by the first component's having changed drastically (Figure 4-10f). You need to be aware of this scenario while using PCA, especially when dealing with both continuous and categorical variables.

```
model.explainedVariance
#Output: DenseVector([0.1434, 0.0987, 0.0787])
```

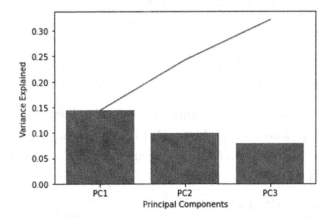

Figure 4-10f. *Variance explained*

Singular Value Decomposition

The best-known method for matrix decomposition is singular value decomposition, or SVD. Any matrix can be decomposed using SVD, thus it is a more stable algorithm. SVD factorizes a matrix into three matrices, U, Σ, and V, such that

$$A = U\Sigma V^{T},$$

where

- A is the original matrix with $r \times c$ dimension. r – rows, c – columns

- U is an $r \times r$ matrix also called as the left singular vectors of A

- Σ is an $r \times c$ diagonal matrix. The diagonal values are the singular values of original matrix A.

- V^{T} is a $c \times c$ matrix also called as the right singular vectors of A

Okay, how does this apply to the concept of variable selection? The answer lies in the Σ component. Let's say you have an image with 1,000 features. Using SVD, you can take the top k singular values by accessing the Σ matrix. You pick this k depending upon the amount of information you want to be represented in the lower dimension. In a PCA scenario, you can think of the variance explained. Let's assume the top 100 (k) out of the 1,000 features explain 80 percent of the information in the data, and you are okay with that. Then, you will pick the top 100 singular values from Σ and substitute the rest of the values with 0 in the Σ matrix. The resulting low-rank matrix can then be used to create a representation of the original data, as shown below.

- U – r x k left singular values

- Σ – k x k singular values

- V^T – c x k right singular values

SVDs are used more for image data compression and in image recovery. Because of the matrix factorization, it also plays a huge role in decomposing a user x item matrix in a collaborative filtering recommendation system. Let's go ahead and look at the PySpark code to implement SVD in our data:

```
from pyspark.mllib.linalg import Vectors
from pyspark.mllib.linalg.distributed import RowMatrix

# convert DataFrame to rdd
df_svd_vector = df.rdd.map(lambda x: x['features'].toArray())
mat = RowMatrix(df_svd_vector)

# Compute the top 5 singular values and corresponding singular vectors.
svd = mat.computeSVD(5, computeU=True)
U = svd.U        # The U factor is a RowMatrix.
s = svd.s        # The singular values are stored in a local dense vector.
V = svd.V        # The V factor is a local dense matrix.
```

You can read more about eigenvalue decomposition of a matrix to get a more in-depth understanding of SVD. Now, let us move on to target-based variable reduction.

Built-in Variable Selection Process: With Target

In the below section we will explore supervised variable selection techniques.

ChiSq Selector

ChiSq stands for *Chi-Square*. It is also called a "goodness of fit" statistic and is denoted by the symbol $\chi 2$. It uses the *Chi-Squared test of independence* to select the best features. This selection process is useful when dealing with categorical variables. However, if you prefer to use continuous variables, you need to bin them into groups before using this technique. Let us understand Chi-Square in detail using the Titanic dataset (https://www.kaggle.com/c/titanic) example that follows.

Gender		Survived		
		No	Yes	Total
Gender	Female	81	231	312
	Male	468	109	577
	Total	549	340	889

Figure 4-11. *Titanic example: gender versus survived*

The table shown in Figure 4-11 is also called a *contingency table*. By looking at the table, we can notice that the percentage of female survivors (231/340 = 67.9%) is much higher than the percentage of male survivors (109/340 = 32.0%). Does this mean there is a relationship between *Gender* and *Survived*? Well, you can answer this question using a Chi-Square test. In a Chi-Square test, you have the following: null and alternate hypotheses.:

> *Null Hypothesis: There is no relationship between the two variables*

> *Alternate Hypothesis: There is a relationship between the two variables*

We will use a 5 percent significance level for our test. The formula for chi-square is provided here:

$$\chi^2 = \sum \frac{(O-E)^2}{E}$$

χ2 – Chi Square, O – Observed data, E – Expected data

Observed data refers to the data that we see in the earlier table. Expected data is calculated based on the totals. The expected value for all four cells is provided in Table 4-3.

Table 4-3. *Expected Values*

Gender = Female and Survived = No	(312*549)/889 = 192.67
Gender = Female and Survived = Yes	(312*340)/889 = 119.32
Gender = Male and Survived = No	(577*549)/889 = 356.32
Gender = Male and Survived = Yes	(577*340)/889 = 220.67

117

Our data looks like that in Table 4-4. We calculated chi-square using the formula shown earlier and summed the *chi-square* column to arrive at the total chi-square value.

Table 4-4. *Chi-Square Value*

Cell	Observed	Expected	Chi-Square
Gender = Female and Survived = No	81	192.67	64.72
Gender = Female and Survived = Yes	231	119.32	104.51
Gender = Male and Survived = No	468	356.32	34.99
Gender = Male and Survived = Yes	109	220.67	56.51
		Total	260.75

The last step is to calculate the degrees of freedom, which is done using the following formula:

```
degrees_of_freedom or df = (r - 1) * (c - 1)
```

where

```
r = number of rows
c = number of columns
```

In our example, $r = 2$ (Gender has two rows) and $c=2$ (Survived has two columns). Therefore, the degrees of freedom is 1. Okay, we are all set. Let us go ahead use the chi-square table (`https://people.richland.edu/james/lecture/m170/tbl-chi.html`) or online chi-square calculator (`https://stattrek.com/online-calculator/chi-square.aspx`) to calculate the p-value. The p-value for this experiment is close to 0. It means that our results are significant, which means that we can reject null hypothesis. Thus, we can conclude that there is a relationship between *Gender* and *Survived*.

Note   Chi-square is susceptible to small frequencies. When the expected value in a cell is less than 5, it can lead to errors in conclusion.

How is this all relevant in variable selection? In our example, we calculated a chi-square value of 260.75, which tells the relationship between two variables (*Gender and Survived*). When the relationship is high, the chi-square value will be high and vice-versa. In a variable selection setting, we can pick variables with high chi-square values because they have a high relationship with the target and thus improve our prediction results.

Note   Chi-square does not measure the strength of relationship. You should use correlation to determine the strength of the relationship.

The code to perform chi-square–based selection is provided next. You might have to rerun assembled vectors because we are only dealing with categorical variables for this type of selection.

```
#select only categorical features present in the dataset
features_list = char_vars #this option is used only for ChiSqselector
#run assembled vectors before running this code with the updated features
list

from pyspark.ml.feature import ChiSqSelector
from pyspark.ml.linalg import Vectors

#chisqselector instance initiation
selector = ChiSqSelector(numTopFeatures=6, featuresCol="features",
                         outputCol="selectedFeatures", labelCol="y")
#fit on the data
chi_selector = selector.fit(df)
#result datasets
result = chi_selector.transform(df)
#outputs
print("ChiSqSelector output with top %d features selected" % selector.
getNumTopFeatures())
print("Selected Indices: ", chi_selector.selectedFeatures)
features_df['chisq_importance'] = features_df['idx'].apply(lambda x: 1 if x
in chi_selector.selectedFeatures else 0)
print(features_df)
```

We have successfully implemented a chi-square–based feature-selection technique for categorical variables.

Model-based Feature Selection

Model-based feature selection requires us to build a machine learning model to find the important features. We will talk about these models in detail in the next chapter. For now, imagine you want a computer to distinguish between *dogs* and *cats*. It is easy for a human to spot and tell the difference, but it is tough for a computer to learn this on its own. That's where a machine learning model comes into the picture. The model teaches the computer to look for specific information to distinguish between *dogs* and *cats*. This specific information is our important features.

In PySpark, tree-based methods (Decision Tree, Random Forest, and Gradient Boosting) provide this option by default. In this section, we will specifically focus on the feature-importance aspect rather than the actual model. The following code is used to calculate the feature importance. We used Random Forest in the bank dataset example. You can replicate the same with other algorithms.

```
from pyspark.ml.classification import RandomForestClassifier
rf = RandomForestClassifier(featuresCol='features', labelCol=target_
variable_name)
rf_model = rf.fit(df)
rf_model.featureImportances
```

   #Output:

```
SparseVector(16, {0: 0.0381, 1: 0.0088, 2: 0.0048, 3: 0.4557, 4: 0.002, 5:
0.0347, 6: 0.0313, 7: 0.0021, 8: 0.0019, 9: 0.001, 10: 0.0001, 11: 0.0112,
12: 0.0017, 13: 0.0215, 14: 0.1484, 15: 0.2366})
```

That's it. We now have our feature importance scores. Let's use the following code to format the output. Remember the `features_df` DataFrame we created in the vector assembler section? We will use it here.

```
import pandas as pd
for k, v in df.schema["features"].metadata["ml_attr"]["attrs"].items():
    features_df = pd.DataFrame(v)
```

```
#temporary output rf_output
rf_output = rf_model.featureImportances
features_df['Importance'] = features_df['idx'].apply(lambda x: rf_output[x]
if x in rf_output.indices else 0)
```

```
#sort values based on descending importance feature
features_df.sort_values("Importance", ascending=False, inplace=True)
```

Now, we have a nice-looking table, as shown in Figure 4-12. The most important feature is *duration*, followed by *poutcome* and so on.

	idx	name	Importance
3	3	duration	0.439098
15	15	poutcome	0.265681
14	14	month	0.139936
5	5	pdays	0.046669
13	13	contact	0.030492
11	11	housing	0.027726
6	6	previous	0.012508
0	0	age	0.012080
1	1	balance	0.006088
2	2	day	0.005995
7	7	job	0.004189
8	8	marital	0.003499
12	12	loan	0.002518
4	4	campaign	0.002192
9	9	education	0.001166
10	10	default	0.000161

Figure 4-12. *Random Forest: feature importance*

In addition, you can use this table to create a feature importance plot, as shown in Figure 4-13.

```
import matplotlib.pyplot as plt

#just for plotting purposes sort it to ascending
features_df.sort_values("Importance", ascending=True, inplace=True)
plt.barh(features_df['name'], features_df['Importance'])
plt.title("Feature Importance Plot")
plt.xlabel("Importance Score")
plt.ylabel("Variable Importance")
```

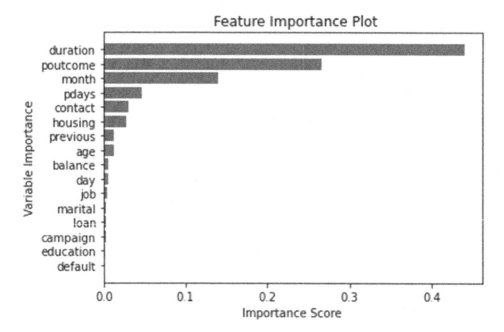

Figure 4-13. *Feature importance plot*

Pretty cool, right? You can use other algorithms to accomplish the same task. However, when dealing with large datasets, Random Forest–based feature selection makes the process a lot easier. One drawback in using tree-based methods is that it uses the training dataset to calculate the feature importance. This can inflate the importance of numeric and high-cardinality variables. To overcome this, you can use permutation importance, which uses the test dataset to calculate the feature importance. Permutation importance is out of scope for this book.

Note   When you use linear models like logistic regression, use the coefficients from the model to showcase feature importance. The higher the **absolute value** of the coefficients, the more important the variable and vice-versa.

Lastly, we would like to focus on *continuous* and *multinomial* targets.

1) For continuous targets, you can use regression trees (RandomForestRegressor or other trees), which can be accessed from the pyspark.ml.regression module. You have to use the LinearRegression module instead of LogisticRegression for a continuous model. The rest of the code piece just discussed can be reused to calculate feature importance.

2) For multinomial targets, you can still use the tree-based models discussed in this section, with the exception of linear model LogisticRegression. It is a bit tricky to calculate feature importance using logistic regression when dealing with multinomial targets. We recommend you stick to tree-based models for such targets.

Until now, we have gone through the code without looking into the formula behind the calculation of feature importance. No worries. We will revisit this section again when we talk about trees in the next chapter.

EXERCISE 4-2: MODEL-BASED SELECTION

You can use either the bank dataset or the housing dataset.

Question 1: Implement decision tree feature importance. Compare and contrast with Random Forest output.

Question 2: Implement gradient-boosted tree feature importance. Compare and contrast with Random Forest output.

Question 3: Implement logistic regression feature importance. Compare and contrast with Random Forest output. (Hint: Use the coefficient of logistic regression. Note: Logistic regression produces dense vector instead of sparse vector.)

Custom-built Variable Selection Process

In the section below we will introduce some custom variable selection techniques.

Information Value Using Weight of Evidence

Weight of evidence (WOE) and information value (IV) are simple yet powerful techniques to perform *variable transformation and selection.* These concepts are closely connected with the logistic regression modeling technique. It is widely used in credit scoring to measure the separation of good versus bad customers. For example, if you are trying to predict whether customers will default on a payment, all the customers who have defaulted will be considered events and those who haven't can be considered non-events.

The formula to calculate WOE and IV is provided here:

$$WOE = \ln\left(\frac{Event\%}{NonEvent\%}\right)$$

$$IV = \sum\left(Event\% - NonEvent\%\right) * \ln\left(\frac{Event\%}{NonEvent\%}\right)$$

Or simply,

$$IV = \sum\left(Event\% - NonEvent\%\right) * \left(WOE\right)$$

Figure 4-14 shows how to calculate these values.

Variable Name	Min. Value	Max. Value	Count	# Event	# Non Event	Event%	Non event%	WOE	Event% - Non event%	IV
Age	10	20	1200	150	1050	28.3%	19.0%	0.3992	9.3%	0.03718
Age	21	30	900	120	780	22.6%	14.1%	0.4733	8.5%	0.04040
Age	31	40	1090	110	980	20.8%	17.7%	0.1580	3.0%	0.00479
Age	41	50	1460	100	1360	18.9%	24.6%	-0.2650	-5.7%	0.01517
Age	50	inf	1410	50	1360	9.4%	24.6%	-0.9582	-15.2%	0.14525
Total			6060	530	5530					0.24279

Figure 4-14. *WOE and IV calculation*

The advantages of WOE transformation are as follows:

1. Handles missing values

2. Handles outliers

3. The transformation is based on logarithmic value of distributions. This is aligned with the logistic regression output function.

4. No need for dummy variables

5. By using proper binning technique, it can establish a monotonic relationship (either increase or decrease) between the independent and dependent variables.

In addition, the IV value can be used to select variables based on the predictive power table shown in Figure 4-15.

Information Value (IV)	Predictive Power
< 0.02	useless for prediction
0.02 to 0.1	weak predictor
0.1 to 0.3	medium predictor
0.3 to 0.5	strong predictor
> 0.5	suspicious or too good to be true

Figure 4-15. *IV and predictive power*

As shown in Figure 4-14 for WOE and IV calculation, you need categories to perform the calculation. Now, you may be wondering, what should I do with continuous variables? As we mentioned before, a proper binning technique should be used to convert continuous to categorical before performing WOE and IV calculations. When the number of continuous variables is small (say, less than ten), you can create your own custom bins and use them for WOE calculations. Even then, it is a tedious task to create the custom bins. Now, think about a large dataset with hundreds of continuous variables. It would be a nightmare to create custom bins. That's where monotonic binning comes into play.

Monotonic Binning Using Spearman Correlation

First, let us understand what is meant by monotonic.

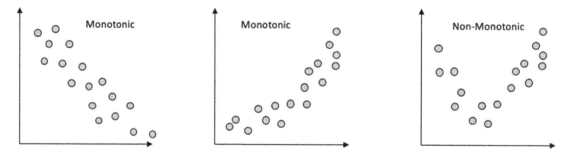

Figure 4-16. *Monotonic versus non-monotonic*

As you can see from Figure 4-16, in the first two scenarios, as the x-axis variable increases, the y-axis value has a single trend: either increasing or decreasing. This is a monotonic relationship. In the third scenario, the x-axis variable change does not have a single trend: either increasing or decreasing behavior. Therefore, it is non-monotonic. A monotonic relationship is useful when it comes to explanatory purposes. When two variables exhibit a monotonic relationship, you can easily explain the results to a non-technical audience. Let's say, for example, the balance and target variables have a relationship, as shown in Figure 4-17.

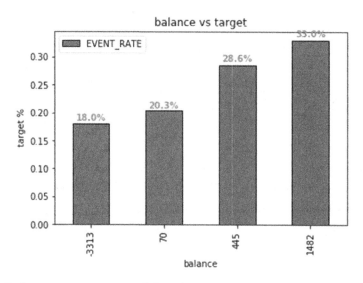

Figure 4-17. *Balance versus target bins*

By looking at this graph, you can easily say, "As the balance increases, the target conversion rate increases." This result is easy to interpret for a non-technical audience.

To perform monotonic binning, we need to understand the Spearman correlation. The Spearman correlation measures the strength and the direction of the monotonic association between two variables. Okay, how does it differ from the Pearson correlation? In a Pearson correlation, we test for a linear relationship between two variables. In this case, the second graph in Figure 4-16 would have a moderate linear relationship as per the Pearson correlation. However, the Spearman correlation would consider it a strong monotonic relationship. The Spearman correlation is also known as the *rank-order correlation*.

How Do You Calculate the Spearman Correlation by Hand?

The formula for the Spearman correlation is provided here:

$$\rho = 1 - \frac{6\sum d_i^2}{n(n^2 - 1)}$$

Here, ρ is the Spearman correlation, d_i is the difference in paired ranks, and n is the number of cases. When you have tied ranks, you need to use the following formula. Rank is the ordered number in a result set.

$$\rho = \frac{\sum_i (x_i - \bar{x})(y_i - \bar{y})}{\sqrt{\sum_i (x_i - \bar{x})^2 \sum_i (y_i - \bar{y})^2}}$$

Here, i is the paired score. For now, we will use only the first formula to calculate the Spearman correlation (Figure 4-18).

Mathematics (mark)	Physics (mark)	Rank for Mathematics	Rank for Physics	d_i	d_i^2
48	56	7	7	0	0
75	89	3	2	1	1
53	65	6	6	0	0
68	72	5	5	0	0
92	74	1	4	-3	9
88	81	2	3	-1	1
74	99	4	1	3	9
				Total	20

Figure 4-18. *Spearman correlation calculation*

Using d_i^2 as 20 and n as 7, we can calculate the ρ value using the first equation:

```
p = 1 - ((6*20)/(7*((7*7) - 1))) = 0.64285714
```

What do we infer from this number 0.64? In general, a Spearman correlation value ranges from -1 to +1. When the value is close to 1 or -1, the monotonic relationship between the two variables is stronger. When the value is close to 0, there is no monotonic relationship between the two variables. In our example, mathematics and physics grades exhibit a strong monotonic relationship. Figure 4-19 is a guide to use to determine the strength of the relationship.

Absolute ρ value	Relationship
0.00 to 0.19	very weak
0.20 to 0.39	weak
0.40 to 0.59	moderate
0.60 to 0.79	strong
0.80 to 1.0	very strong

Figure 4-19. *Spearman correlation value guide*

How Is Spearman Correlation Used to Create Monotonic Bins for Continuous Variables?

In the code for monotonic binning, you will notice that the mono_bin() function uses a max_bin option. This is a user-specified value. For continuous variables, we start with the max_bin provided by the user and decrease the bins as each step proceeds. We create quantiles based on the max_bin. Next, we calculate the Spearman correlation based on the newly created quantile bins and the target variable. Until we establish a strong monotonic relationship between the continuous variable and the target variable, we keep on decreasing the max_bin value to create new quantiles and calculate the Spearman value. When we reach a strong correlation between the two variables, the binning process is stopped.

Here is the complete code to calculate WOE and IV, including the monotonic binning option for continuous variables. We have provided the entire code from data load to WOE and IV calculation. This code can be used on any dataset. And, the code is provided as a self-study.

```python
# Default parameters
from pyspark.sql import SparkSession
spark = SparkSession.builder.getOrCreate()
dataset = 2 # 1 or 2
# Load Dataset
from pyspark.sql import functions as F

# if the option is set as 1 load bank dataset
if dataset == 1:
    filename = "bank-full.csv"
    target_variable_name = 'y'
    df = spark.read.csv(filename, header=True, inferSchema=True, sep=';')
    df = df.withColumn(target_variable_name, F.when(df[target_variable_
    name] == 'no', 0).otherwise(1))
# if the option is not set as 1 load housing dataset
else:
    filename = "melb_data.csv"
    target_variable_name = "type"
    df = spark.read.csv(filename, header=True, inferSchema=True, sep=',')
    df = df.withColumn(target_variable_name, F.when(df[target_variable_
    name] == 'h', 0).otherwise(1))

df.show()
# target check
df.groupby(target_variable_name).count().show()
#identify variable types and perform some operations
def variable_type(df):

    vars_list = df.dtypes
    char_vars = []
    num_vars = []
    for i in vars_list:
```

```
        if i[1] in ('string'):
            char_vars.append(i[0])
        else:
            num_vars.append(i[0])

    return char_vars, num_vars

char_vars, num_vars = variable_type(df)

if dataset != 1:
    char_vars.remove('Address')
    char_vars.remove('SellerG')
    char_vars.remove('Date')
    char_vars.remove('Suburb')
num_vars.remove(target_variable_name)
final_vars = char_vars + num_vars

# WOE & IV code
from pyspark.sql import functions as F
import pandas as pd
import numpy as np
from pyspark.ml.feature import QuantileDiscretizer
from pyspark.ml.feature import VectorAssembler
import scipy.stats.stats as stats

#default parameters
#rho value for Spearman correlation. You can adjust this value
custom_rho = 1
#maximum number of bins to start of with. It will keep decreasing from this
number. Adjustable parameter
max_bin = 20

# This function calculates the WOE and IV values based on the PySpark
output.
# Note: This portion of the code is Python-based implementation for
efficieny purposes
def calculate_woe(count_df, event_df, min_value, max_value, feature):

    # implemeting the table structure shown in WOE & IV calculation figure
```

```
    woe_df = pd.merge(left=count_df, right=event_df)
    woe_df['min_value'] = min_value
    woe_df['max_value'] = max_value
    woe_df['non_event'] = woe_df['count'] - woe_df['event']
    woe_df['event_rate'] = woe_df['event']/woe_df['count']
    woe_df['nonevent_rate'] = woe_df['non_event']/woe_df['count']
    woe_df['dist_event'] = woe_df['event']/woe_df['event'].sum()
    woe_df['dist_nonevent'] = woe_df['non_event']/woe_df['non_event'].sum()
    woe_df['woe'] = np.log(woe_df['dist_event']/woe_df['dist_nonevent'])
    woe_df['iv'] = (woe_df['dist_event'] - woe_df['dist_nonevent'])*woe_
                                df['woe']
    woe_df['varname'] = [feature]* len(woe_df)
    woe_df = woe_df[['varname','min_value', 'max_value', 'count',
'event', 'non_event', 'event_rate', 'nonevent_rate', 'dist_event','dist_
nonevent','woe', 'iv']]
    woe_df = woe_df.replace([np.inf, -np.inf], 0)
    woe_df['iv'] = woe_df['iv'].sum()
    return woe_df

#monotonic binning function implemented along with Spearman correlation
def mono_bin(temp_df, feature, target, n = max_bin):

    r = 0
    while np.abs(r) < custom_rho and n > 1:

        try:
            #Quantile discretizer cuts data into equal number of
             observations
            qds = QuantileDiscretizer(numBuckets=n, inputCol=feature,
                    outputCol='buckets', relativeError=0.01)
            bucketizer = qds.fit(temp_df)
            temp_df = bucketizer.transform(temp_df)

            #create corr_df is Python-based implementation for efficiency
             purposes
            corr_df = temp_df.groupby('buckets').agg({feature:'avg',
                    target:'avg'}).toPandas()
```

131

```
            corr_df.columns = ['buckets', feature, target]
            r, p = stats.spearmanr(corr_df[feature], corr_df[target])
                    #spearman correlation
            n = n - 1
        except Exception as e:
            n = n - 1

        return temp_df


#execute WOE for all the variables in the dataset
def execute_woe(df, target):

    count = -1
    for feature in final_vars:
        #execute if the feature is not a target column name. Provided as an
         extra check.
        if feature != target:
            count = count + 1
            temp_df = df.select([feature, target])

            #perform monotonic binning for numeric variables before doing
             WOE calculation
            if feature in num_vars:
                temp_df = mono_bin(temp_df, feature, target, n = max_bin)
                # group buckets in numerical
                grouped = temp_df.groupby('buckets')
            else:
                # just group categories in categorical
                grouped = temp_df.groupby(feature)

            #count and event value for each group
            count_df = grouped.agg(F.count(target).alias('count')).
            toPandas()
            event_df = grouped.agg(F.sum(target).alias('event')).toPandas()

            #store min and max values for variables. for category both
             takes the same value.
            if feature in num_vars:
```

```
        min_value = grouped.agg(F.min(feature).alias('min')).
        toPandas()['min']
        max_value = grouped.agg(F.max(feature).alias('max')).
        toPandas()['max']
    else:
        min_value = count_df[feature]
        max_value = count_df[feature]

    #calculate WOE and IV
    temp_woe_df = calculate_woe(count_df, event_df, min_value, max_
                value, feature)

    #final dataset creation
    if count == 0:
        final_woe_df = temp_woe_df
    else:
        final_woe_df = final_woe_df.append(temp_woe_df, ignore_
                index=True)

    # separate IV dataset creation
    iv = pd.DataFrame({'IV':final_woe_df.groupby('varname').iv.max()})
    iv = iv.reset_index()
    return final_woe_df, iv

# invoke WOE & IV code
output, iv = execute_woe(df, target_variable_name)
```

That is all. We are done with weight of evidence and information value.

Custom Transformers

So far, we have been looking at some high-level APIs available in Spark ML that can be used to perform variable selection (built-in selection techniques discussed earlier). In this section, we will create a custom Spark ML transformer to accomplish the same thing. Why do we need to discuss custom transformers?

Take a moment to think about the RandomForest code that we used for variable selection earlier:

```
from pyspark.ml.classification import RandomForestClassifier
rf = RandomForestClassifier(featuresCol='features', labelCol=target_
variable_name)
rf_model = rf.fit(df)
rf_model.featureImportances
```

With a few lines of code, we did the following: (1) invoke the RandomForest; (2) define an instance of RandomForest object; (3) fit the model on your dataset; and (4) show the importance score. This is the power of object-oriented programming (OOP). Here is a quick refresher on OOP concepts:

1. *Class* – A user-defined blueprint that is used to create object instance. In code line 2, RandomForestClassifier is a class and rf is the object instance.

2. *Methods* – A function written inside a class is called a method. In line 3, the fit function is a *method*.

3. *Abstraction* – This is a concept by which only the essential information is shown to the users. This is called implementation hiding. In the RandomForestClassifier code shown earlier, we only provide the input variable, target, and our dataset to fit the model. We don't see any details on how the Random Forest model is implemented in the background.

4. *Encapsulation* – This is a concept in which you hide information from users. This is called information hiding. Any *variable/ method* can be hidden using this technique. The *variable/ method* can be *protected* (single underscore in the front, cannot be accessed outside a class, accessible within the class and its subclass), or private (double underscore in the front, accessible only inside the class). When you look at the source code, you might still be able to see the variables/methods. So, they are conceptually hidden, but not visually.

5. *Inheritance* – This is a concept in which you inherit the characteristics of a parent class to create a child class. Think of a square object. It is technically a rectangle with all four sides equal. In this case, rectangle is the parent class and square is the child class.

6. *Polymorphism* – This is a concept in which you can morph the characteristics of an object. It means that you can make the same object do multiple things other than the usual. Think of the *addition (+)* symbol. Originally, it can sum only numeric variables. In Python, when you include addition between strings, it concatenates them. When you execute this code—`print("Hello " + "World")`—it concatenates the string to produce the `Hello World` output. In this case, the *addition (+)* symbol can add strings as well as numerics. This is an example of polymorphism.

Using OOP concepts, you can create a powerful machine learning class object. This makes it simple for end users to use your code. In addition, these objects are easier to combine into a single pipeline or workflow, thereby creating an efficient end-to-end process. Okay, let us quickly look at the main concepts used to create such custom code.

Main Concepts in Pipelines

DataFrame: A collection of different datatypes organized in a single ML dataset

Estimator: This is used to *fit* a function to a DataFrame. In `RandomForestClassifier`, the `fit` function is our estimator.

Transformer: This is used to *transform* DataFrame from one to another. Sometimes, a transformer requires a function to be fit before it can be applied. In `RandomForestClassifier`, the `predict` function is a transformer.

Pipeline: An ML workflow or sequence of steps to be executed to get to the end result.

Parameter: The estimator and transformer use uniform API for specifying parameters.

Let us create a custom correlation `class` and use it for variable selection. Finally, we will demonstrate the pipeline use case. Before we do any implementation, let's quickly go through the code used to calculate correlation on a dataset.

```
from pyspark.mllib.stat import Statistics
import pandas as pd
```

```
correlation_type = 'pearson' # 'pearson', 'spearman'

#transformer function
for k, v in df.schema["features"].metadata["ml_attr"]["attrs"].items():
    features_df = pd.DataFrame(v)
column_names = list(features_df['name'])
df_vector = df.rdd.map(lambda x: x['features'].toArray())
matrix = Statistics.corr(df_vector, method=correlation_type)
corr_df = pd.DataFrame(matrix, columns=column_names, index=column_names)
```

The preceding code produces a pandas DataFrame with the correlation results. We will add some more code to extend its functionality. We will include an option to find the most correlated variables using an external parameter.

```
#transformer continuation
final_corr_df = pd.DataFrame(corr_df.abs().unstack().sort_
values(kind='quicksort')).reset_index()
final_corr_df.rename({'level_0': 'col1', 'level_1': 'col2', 0:
'correlation_value'}, axis=1, inplace=True)
final_corr_df = final_corr_df[final_corr_df['col1'] != final_corr_
df['col2']]
correlation_cutoff = 0.65 #custom parameter
final_corr_df[final_corr_df['correlation_value'] > correlation_cutoff]
```

We recommend you get the feel of the preceding code before you use the custom class.

```
# Import the estimator and transformer classes
from pyspark.ml import Transformer
# Parameter sharing class. We will use this for input column
from pyspark.ml.param.shared import HasInputCol
# Statistics class to calculate correlation
from pyspark.mllib.stat import Statistics
import pandas as pd
# custom class definition
class CustomCorrelation(Transformer, HasInputCol):
    """

    A custom function to calculate the correlation between two variables.
```

Parameters:
----
inputCol: default value (None)
    Feature column name to be used for the correlation purpose. The
    input column should be assembled vector.

correlation_type: 'pearson' or 'spearman'

correlation_cutoff: float, default value (0.7), accepted values 0 to 1
    Columns more than the specified cutoff will be displayed in the
    output dataframe.
    """

```
# Initialize parameters for the function
def __init__(self, inputCol=None, correlation_type='pearson',
correlation_cutoff=0.7):

    super(CustomCorrelation, self).__init__()

    assert inputCol, "Please provide a assembled feature column name"
    #self.inputCol is class parameter
    self.inputCol = inputCol

    assert correlation_type == 'pearson' or correlation_type ==
'spearman', "Please provide a valid option for correlation type. 'pearson'
or 'spearman'. "
        #self.correlation_type is class parameter
        self.correlation_type = correlation_type

    assert 0.0 <= correlation_cutoff <= 1.0, "Provide a valid value for
    cutoff. Accepted range is 0 to 1"
    #self.correlation_cutoff is class parameter
    self.correlation_cutoff = correlation_cutoff

# Transformer function, method inside a class, '_transform' - protected
  parameter
def _transform(self, df):

    for k, v in df.schema[self.inputCol].metadata["ml_attr"]["attrs"].
    items():
        features_df = pd.DataFrame(v)
```

```
column_names = list(features_df['name'])
df_vector = df.rdd.map(lambda x: x[self.inputCol].toArray())

#self.correlation_type is class parameter
matrix = Statistics.corr(df_vector, method=self.correlation_type)

# apply pandas dataframe operation on the fit output
corr_df = pd.DataFrame(matrix, columns=column_names, index=column_
names)
final_corr_df = pd.DataFrame(corr_df.abs().unstack().sort_
values(kind='quicksort')).reset_index()
final_corr_df.rename({'level_0': 'col1', 'level_1': 'col2', 0:
'correlation_value'}, axis=1, inplace=True)
final_corr_df = final_corr_df[final_corr_df['col1'] != final_corr_
df['col2']]

#shortlisted DataFrame based on custom cutoff
shortlisted_corr_df = final_corr_df[final_corr_df['correlation_
value'] > self.correlation_cutoff]
return corr_df, shortlisted_corr_df
```

Voila! We now have our custom correlation class that works similarly to any Spark ML object. This code can be saved in a separate Python file *customcorrelation.py*. Let's test our new class.

```
from customcorrelation import CustomCorrelation
clf = CustomCorrelation(inputCol='features')
output, shorlisted_output = clf.transform(df)
```

That's all. You used three lines of code to calculate correlation among the input features in your dataset. This is similar to a RandomForestClassifier invocation we saw before. Some folks might question whether we can do the same with *user-defined functions (udfs)*. You are right in terms of abstraction of code; however, *udfs* fail to fit in the pipeline workflow. All these efforts we made so far are to make our work pipeline compatible. So, let's test that feature too. To demonstrate this feature, we will combine the assemble vector step and correlation step into one pipeline. If you need to re-run, you will need to execute till *"Step 2: Apply StringIndexer to character columns"* discussed earlier in this chapter.

```
from pyspark.ml.feature import VectorAssembler
from customcorrelation import CustomCorrelation
from pyspark.ml import Pipeline

#exclude target variable and select all other feature vectors
features_list = df.columns
features_list.remove(target_variable_name)
stages = []

#assemble vectors
assembler = VectorAssembler(inputCols=features_list, outputCol='features')
custom_corr = CustomCorrelation(inputCol=assembler.getOutputCol())
stages = [assembler, custom_corr]

#use pipeline to process sequentially
pipeline = Pipeline(stages=stages)

#pipeline model
pipelineModel = pipeline.fit(df)

#apply pipeline model on data
output, shorlisted_output = pipelineModel.transform(df)
```

Pretty cool! Now you can implement your own custom transformers and create a neat ML workflow.

Voting-based Selection

So far, we have discussed multiple techniques for the variable selection process. Wouldn't it be great to combine all the variable selection options into one process? We will accomplish this using a voting-based schema. The idea is to apply multiple techniques to select variables. When a variable selection technique picks a variable, we assign a score to the variable. In other words, the technique voted for that variable. At the end, we calculate the total score for each variable and select the variables with the highest scores (Figure 4-20). This way, we end up picking the variables with the highest "preference" from a variety of selection techniques, which, depending on the case at hand, might not work well when used individually.

Variable Name	Random Forest	Decision Tree	Chi Square	Gradient Boosting	WOE & IV	Total Votes
Age	1	1	1	0	1	4
Gender	1	1	0	1	1	4
Income	1	0	0	1	1	3
Education	0	0	1	1	0	2
Occupation	1	0	0	0	0	1
Homeowner	0	0	0	0	0	0

Figure 4-20. *Voting-based selection*

This section requires you to complete *Exercise 4-2: Model-based selection*. For the sake of demonstration, we have provided the solution for the exercise here (Figure 4-21). We will use all the target-based selection techniques we discussed earlier in this chapter.

```
#decision tree algorithm - solution to Question 1
from pyspark.ml.classification import DecisionTreeClassifier
dt = DecisionTreeClassifier(featuresCol='features', labelCol=target_
variable_name)
dt_model = dt.fit(df)
dt_output = dt_model.featureImportances
features_df['Decision_Tree'] = features_df['idx'].apply(lambda x: dt_
output[x] if x in dt_output.indices else 0)

#Gradient boosting algorithm - solution to Question 2
from pyspark.ml.classification import GBTClassifier
gb = GBTClassifier(featuresCol='features', labelCol=target_variable_name)
gb_model = gb.fit(df)
gb_output = gb_model.featureImportances
features_df['Gradient Boosting'] = features_df['idx'].apply(lambda x: gb_
output[x] if x in gb_output.indices else 0)

#logistic regression algorithm - solution to Question 3
```

```
from pyspark.ml.classification import LogisticRegression
lr = LogisticRegression(featuresCol='features', labelCol=target_variable_
name)
lr_model = lr.fit(df)
lr_output = lr_model.coefficients
#absolute value is used to convert the negative coefficients. This should
be done only for feature importance.
features_df['Logistic Regression'] = features_df['idx'].apply(lambda x:
abs(lr_output[x]))

#random forest - technique which we discussed before
from pyspark.ml.classification import RandomForestClassifier
rf = RandomForestClassifier(featuresCol='features', labelCol=target_
variable_name)
rf_model = rf.fit(df)
rf_model.featureImportances
rf_output = rf_model.featureImportances
features_df['Random Forest'] = features_df['idx'].apply(lambda x: rf_
output[x] if x in rf_output.indices else 0)

#voting-based selection
features_df.drop('idx', axis=1, inplace=True)
num_top_features = 7 #top n features from each algorithm
columns = ['Decision_Tree', 'Gradient Boosting', 'Logistic Regression',
'Random Forest']
score_table = pd.DataFrame({},[])
score_table['name'] = features_df['name']
for i in columns:
    score_table[i] = features_df['name'].isin(list(features_
df.nlargest(num_top_features,i)['name'])).astype(int)
score_table['final_score'] = score_table.sum(axis=1)
score_table.sort_values('final_score',ascending=0)
```

	name	Decision_Tree	Gradient Boosting	Logistic Regression	Random Forest	final_score
14	month	1	1	1	1	4
15	poutcome	1	0	1	1	3
3	duration	1	1	0	1	3
0	age	1	1	0	1	3
11	housing	0	1	1	1	3
13	contact	1	1	1	0	3
6	previous	1	0	0	1	2
5	pdays	0	1	0	1	2
1	balance	1	0	0	0	1
2	day	0	1	0	0	1
12	loan	0	0	1	0	1
8	marital	0	0	1	0	1
10	default	0	0	1	0	1
7	job	0	0	0	0	0
4	campaign	0	0	0	0	0
9	education	0	0	0	0	0

Figure 4-21. *Voting-based selection output*

Summary

- We went through variable selection techniques in detail.

- We discussed target- versus non-target–based selection techniques.

- We created our own Spark ML transformer and fit it with the ML workflow.

- Finally, we combined all the selection techniques to create a vote-based selection technique.

Great job! In the next chapter, we will learn the most exciting part in this book—machine learning models. Keep learning and stay tuned.

CHAPTER 5

Supervised Learning Algorithms

It's time to do some learning based on the data. Most folks think machine learning is applying an algorithm on given data and then predicting results. Well, it's not just that. Eighty percent of the work involves data collection, preprocessing, cleaning, feature engineering, transformation, and selecting the best features. The remaining 20 percent is spent on building machine learning models, validation, and deployment. The entire operation is called MLOps (machine learning operations). It is similar to DevOps, but for machine learning. In order to understand and deploy a production model, you should be familiar with each component in MLOps. In this book, the chapters we have covered so far have discussed the 80 percent of the work. If you skipped those chapters, we recommend reading them before you read this chapter. In addition, business and domain knowledge helps you to improve the process throughout.

Basics

Okay, let's now discuss machine learning algorithms. Broadly, these algorithms are categorized into three types: supervised, unsupervised, and reinforcement learning (Table 5-1).

Table 5-1. *Types of Machine Learning Algorithms*

	Supervised	Unsupervised	Reinforcement
What is it?	Algorithm is trained on labeled data.	Algorithm is trained on unlabeled data.	Agent is trained based on the rewards earned for each action.
Where is it used?	Regression, classification	Clustering, segmentation, association	Rewards-based
What type of data?	Labeled data. Example: Predict cat vs. dog or predict house price	Unlabeled data; just input features. Example: Customer profiling	No predefined data. Example: Self-driving cars, game agent
Is supervision required?	Yes	No	No
What is the goal?	Reduce the error in target prediction using the known input/output mapping	Understand the pattern and behaviors in the data to tell a story	Trial and error method. Make the agent learn based on its errors and to improve rewards in the long run.

Apart from the categories mentioned, there are two other types of algorithms that are common these days: recommendation systems and semi-supervised algorithms. A recommendation system could be either supervised or unsupervised. We will learn more about this in the next chapter. A semi-supervised algorithm combines both supervised and unsupervised learning approaches—a small amount of labeled data to classify the large amount of unlabeled data. Semi-supervised algorithms and reinforcement learning are out of the scope for this book. In this chapter, we will cover supervised learning algorithms. In the next chapter, we will cover unsupervised and recommendation algorithms.

Within supervised algorithms, we have these types: regression and classification (binary, multinomial, multi-label).

Regression

In a regression model, we predict continuous values, such as house price. Regression is a statistical method used in finance, investing, and other disciplines that attempts to determine the strength and character of the relationship between one dependent variable (usually denoted by Y) and a series of other variables (known as independent variables).

Classification

In a classification model, we predict discrete values; for example, cats versus dogs, spam versus not spam, and so on. When the number of categories is two, it is called binary classification. When the number of categories is more than two, it is called multinomial classification. Binary classification models are a subset of multinomial classification models. Another subset of multinomial classification models is ordinal classification. In an ordinal model, the targets are ordered; for example, when predicting the likelihood of purchasing an economy ticket versus a business ticket versus a first-class ticket. An important thing to remember is that the target is mutually exclusive in these classification models. In some cases, there could be multiple targets for a single observation; for example, the curation of articles. This book can be curated into these titles: "Machine learning, Data science, PySpark applications." Any of these three is a valid target. This type of task is a multi-label classification.

In either case (regression or classification), the goal is to reduce the error between the predicted values and the original value. We will start with the goal (reduce error) to cover some related concepts and then dive into algorithms.

Loss Functions

Loss functions are the error functions associated with each supervised learning algorithm (Table 5-2). We try to minimize the loss function in any model-training process. Why minimize it? Let's take an example of predicting the weight of a person. Our original weight values are 30, 40, and 50. Let's assume one of our models predicts the weights as 32, 46, and 57, respectively.

Table 5-2. *Types of Loss Functions*

	Name	Equation	When to use?
Regression	Squared error loss	$(y - \hat{y})^2$	Commonly used loss function in regression. Easily differentiable loss.
	Absolute error loss	$\|y - \hat{y}\|$	Outliers are present in data. Absolute loss is more robust to outliers.
	Pseudo-Huber loss	$x = \|y - \hat{y}\|$ $$\begin{cases} (y - y - hat)^2 \; ; when\ x \le \alpha \\ y - y - hat \; ; otherwise \end{cases}$$	Combines both squared error and absolute error loss. Tune hyperparameter α to get best results.
Classification	Binary cross-entropy loss	$-(y \log(p) + (1-y) \log(1-p))$	Binary targets. Default loss function.
	Categorical cross-entropy loss	C - number of classes $$-\sum_{x \in C} p(x) \log q(x)$$	Multinomial targets. Default loss function.
	KL divergence loss	C - number of classes $$-\sum_{x \in C} p(x) \log \left(\frac{q(x)}{p(x)} \right)$$	Multinomial targets *Note: Cross-entropy = Entropy + KL Divergence*
	Hinge loss	$x = \hat{y} * y$ $\max(0, 1 - x)$	Binary targets. Typically used in support vector machines. Used to maximize margin.

Using the squared error loss function, the total error for this model is as follows:

```
Model 1 error = (30 -32)² + (40 - 46)² + (50 - 57)² = 4 + 36 + 49 = 89
```

A second model predicts the weights as 31, 42, and 52. For this model, the squared error is as follows:

```
Model 2 error = (30 -31)² + (40 - 42)² + (50 - 52)² = 1 + 4 + 4 = 9
```

As you can see, when you start to minimize the squared error function, the prediction starts to come close to the original values. Therefore, understanding the loss function and choosing the appropriate one is an important task in a model-training process. You can visualize the loss function here: `https://academo.org/demos/3d-surface-plotter/`. We showcase an example using the squared error loss in Figure 5-1.

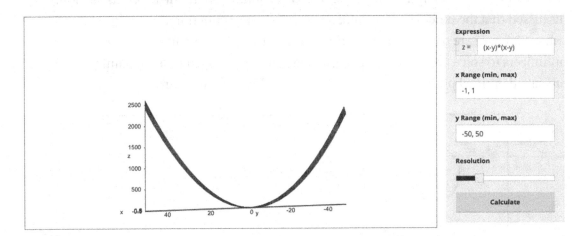

Figure 5-1. *Squared error loss*

When you reach the bottom of the loss function, the squared error value is minimal. In a model-training process, the model starts with an error and takes subsequent steps to reach the bottom value of the curve. This process is called gradient descent. We have shown another loss function—absolute error loss—in Figure 5-2.

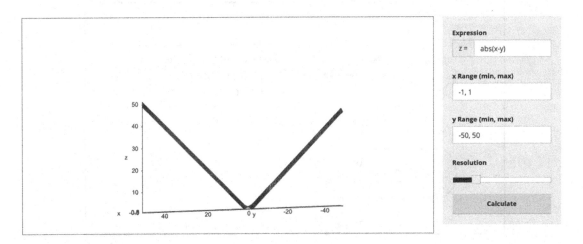

Figure 5-2. *Absolute error loss*

As we explore the algorithms in later sections, we will switch the loss functions depending on the model.

Optimizers

Optimizers are used to find the minimum value of a function. In the previous section, we discussed that the model error is minimal when the loss function reaches the minimum value. Since we are trying to find the minimum value of the loss function, we can use optimizers to solve these tasks. The model parameters that produce the minimal error are called optimal parameters. The whole path is depicted in Figure 5-3.

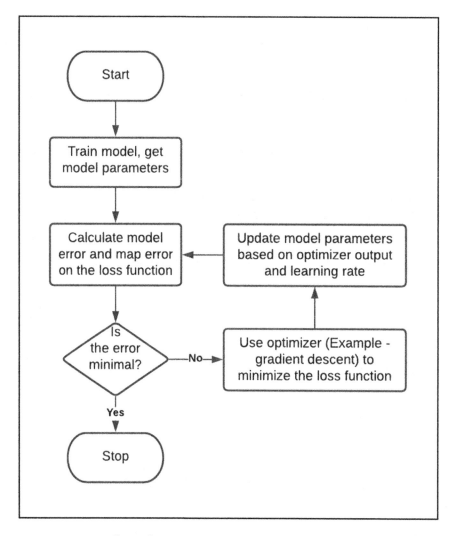

Figure 5-3. *Optimizer flow chart*

In this section, we will deep dive into different optimizers to accomplish this task. Let's start with gradient descent.

Gradient Descent

Imagine you are at the top of the mountain and your goal is to reach the bottom of the mountain using the shortest way possible. You won't jump directly from the top, because you don't know the height of the mountain. It would be crazy; you would most probably die. The unknown knowledge— *"Where is the bottom of the mountain?"*—is the function we are trying to solve here. To find this solution, we start descending the mountain by taking *small steps* in the *right direction*. The small step ensures you can adapt to the terrain, and the right direction ensures you will descend the mountain. You repeat the process (small steps and right direction descent) several times to get to the bottom of the mountain. This whole idea is called gradient descent.

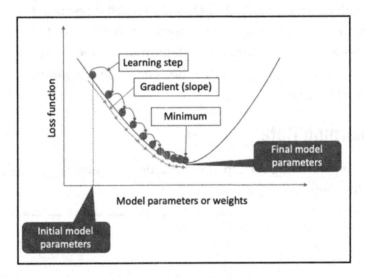

Figure 5-4. *Gradient descent*

Mathematically, the gradient is the slope value of the function that the optimizer is trying to solve. Hence, it is called gradient descent. It is shown using the orange line in Figure 5-4. The model parameters are updated using this equation:

$$\beta_i = \beta_i - \alpha * \text{Slope}_i$$

Where

$\beta_i = \text{Model parameter of a variable } i$

$\alpha = \text{Learning rate}$

$\text{Slope}_i = \text{Gradient or slope with respect to the variable } i = \dfrac{\partial}{\partial \beta_i} \text{Error}$

In the initial phase of model training, the slope is large and starts decreasing as we reach the optimal solution. You can stop the model-training process when one of the following conditions is satisfied:

1. Slope (gradient) reaches close enough to zero. At this point, the model parameter updates yield the same result.

2. The number of iterations or epochs is met.

3. The model improvement falls below a specified threshold. To accomplish this, you compare previous iteration against the current iteration's performance to determine the model (delta) improvement and check whether the delta value is below the threshold specified.

Choosing Learning Rate

We saw the learning rate being used to update model parameters. The question is, "What is the best learning rate?"

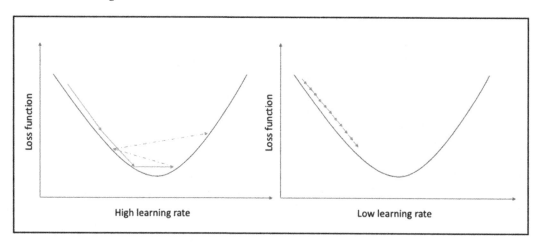

Figure 5-5. *Learning rate*

- By choosing a high learning rate, you descend much faster, and you could potentially miss the bottom of the curve. This is shown using the dotted lines in Figure 5-5. In this case, the model starts to produce chaotic results.

- However, by choosing a low learning rate, your model will take more time to converge on the optimal solution.

When your model is not producing the expected results, decrease the learning rate and rerun it. This is a hyperparameter that can be tuned according to your data and model needs. Another approach is to use an adaptive learning rate. In this case, the learning rate changes according to the slope of the curve. During the initial model training, it starts with a high learning rate (when slope is high), and as it descends the slope, the learning rate can be decreased. This way, you can reach the optimal solution much faster and you can get the best of both worlds (high and low learning rates). The adaption function could be linear or an order of magnitude. We will discuss more about adaptive learning rate in a later section.

Local Minima Versus Global Minima

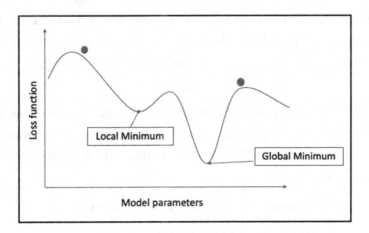

Figure 5-6. *Local minima versus global minima*

In Figure 5-6, we can see that the actual bottom of the curve is the global minimum position. Using gradient descent, you would not always be able to reach the bottom of the curve. Depending upon the initial parameters, the model could start descending from either the left or the right. If it started from the right, it would have a high chance of reaching the global minimum. However, if it started from the left, it could be stuck in the local minimum, and the gradient descent operation would stop. This is a problem with gradient descent. It cannot differentiate between local minimum and global minimum. You can avoid this problem by restarting the model-training process with different hyperparameters.

Model-training Time

As you know by now, the parameter update happens after all the samples in your training data are run through the gradient descent optimization process. For smaller datasets, this approach is okay. However, when you are dealing with millions of pieces of data, gradient descent optimization is very slow. The training time of the model increases drastically as the amount of data increases.

Stochastic/Mini-batch Gradient Descent

To solve the model-training time and the local minimum problem, we will take a look at a variant of gradient descent. In stochastic gradient descent (SGD), the parameter update is done for every observation in the dataset. If your dataset has 100 rows, the parameter update happens 100 times in a single pass. This improves the training time and model convergence when the dataset size is huge. In addition, the randomness introduced by the SGD process helps the model to escape the local minimum and reach the global (or better) minimum. Even though SGD has these advantages, there are some drawbacks too:

- Due to the randomness introduced each time, the optimizer takes noisy steps to converge to the minimum. Sometimes, these noisy steps could throw you off-course, and you could end up in one of the local minimums.

- Frequent SGD updates are computationally expensive.

- SGD cannot perform vectorized updates, since it uses one observation at a time.

Another variant of gradient descent is mini-batch gradient descent. Let's say you have 100 rows in the training data. In a single pass, gradient descent uses all 100 rows to perform one weight update, whereas SGD does a weight update for each observation (so 100 times). Is there a middle ground? Mini-batch gradient descent is the solution to this question. In a mini-batch GD, based on the batch size (let's say, 5), the weight update happens after each batch is processed. In our example, it happens 20 (100 rows /5 batch size = 20) times. An optimal batch size could help the model train and converge faster on the global minimum. Depending on the size of the data, a good default batch size is 32. Clearly, mini-batch GD overcomes the disadvantages listed in SGD. Still, mini-batch GD could sometimes converge on the local minimum instead of the global minimum.

Momentum

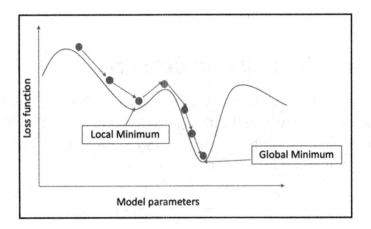

Figure 5-7. *Momentum*

Let's look at the equation for momentum optimizer and then talk about the details (Figure 5-7).

$$\beta_i = \beta_i - \alpha * \text{Slope}_i(t) + m = GD + m$$

$$m = w * (-\alpha * \text{Slope}_i(t-1))$$

Where

i = Model parameter of a variable i

(t) = current update

GD = Gradient descent

m = momentum value

w = weightage provided the previous update (typically 0.9)

$(t-1)$ = previous update

To overcome the local minimum problem, we extend the gradient descent equation by adding a momentum component. You can think of momentum as rolling a ball downhill. The ball gains momentum as it rolls down the hill. Even when the ball hits a small dip along the way (local minimum), it will still keep on rolling until it reaches a flat surface (global minimum). Without momentum, the ball would have rolled with a constant speed and would have ended up in the small dip (local minimum). With momentum, we incorporate the knowledge that the ball is moving downhill. This is done by adding the speed at which the ball was rolling in the previous step. The weightage component (w) decides the amount of weightage to be given for the previous step. The w is a hyperparameter that can be tuned for best results.

AdaGrad (Adaptive Gradient) Optimizer

We can also escape the local minimum by using an adaptive learning rate. We discussed this concept earlier, and we will see the nitty-gritty details in this section. The adaptive gradient is achieved by dividing the learning rate by an exponentially increasing parameter (G_i). This is shown in the following equation:

$$\beta_i = \beta_i - \frac{\propto}{\sqrt{G_i(t) + \varepsilon}} * \text{Slope}_i(t)$$

And G_i is given by

$$G_i(t) = G_i(t-1) + \left(\text{Slope}_i(t)\right)^2$$

At initial model training, $G_i(0) = 0$. The slope term is always positive, and therefore $G_i(t)$ keeps on increasing in each iteration. Since we are dividing the learning rate by an increasing number, the learning rate keeps on decreasing. This is the whole concept behind AdaGrad. The epsilon (ε) parameter is set to a very small value to avoid divide by zero errors.

Root Mean Square Propagation (RMSprop) Optimizer

RMSprop is a variant of the AdaGrad optimizer. Let's look at the following equation:

$$\beta_i = \beta_i - \frac{\propto}{\sqrt{G_i(t)} + \varepsilon} * \text{Slope}_i(t)$$

The update rule is still the same. However, the change is introduced in the G_i parameter, as follows:

$$G_i(t) = \gamma G_i(t-1) + (1-\gamma)\left(\text{Slope}_i(t)\right)^2$$

At initial model training, $G_i(0) = 0$. The difference between RMSprop and AdaGrad comes from the Gamma parameter (γ). This is another hyperparameter that can be optimized. Typically, the value for γ is set to 0.9. The γ parameter introduces a moving average to the gradient component. Since we take the square root of the moving average, this optimizer is called the Root Mean Square optimizer.

Adaptive Moment (Adam) Optimizer

Adaptive Moment combines RMSprop and Momentum and hence the name Adaptive Moment. This can help achieve momentum and a learning rate schedule (adaptive learning rate) by using one optimizer. The updated equation is provided here:

$$\beta_i = \beta_i - \frac{\propto}{\sqrt{G_i(t)} + \varepsilon} * m_i(t)$$

where, $m_i(t)$ is the momentum component and is calculated using the equation

$$m_i(t) = w * m_i(t-1) + (1-w) * \text{Slope}_i(t)$$

At initial model training, $m_i(0) = 0$. w is the weightage parameter that we used in the Momentum equation. Thus, $m_i(t)$ is the first moment estimate and $G_i(t)$ is the second moment estimate.

Quick Recap

- Error functions are also called loss functions.

- When you reach the bottom of a loss function, the model error is at its minimum. At this point, the model fits the data better.

- You can minimize the error function using optimizers such as gradient descent.

- Local minimum and global minimum issues can be solved by adding momentum, tweaking the learning rate, or doing both.

Activation Functions

Another component in an algorithm is the activation function. A simple way to think about the activation function is to look at the biological neurons. When the body needs some energy, the brain activates the food neuron. The person then thinks, "I want to eat food." The brain fires these activations to communicate with the person and make them take action based on it.

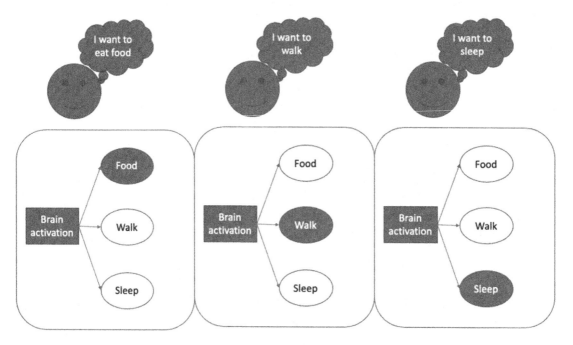

Figure 5-8. *Activation function illustration*

In a similar fashion, when it comes to training an algorithm, activation functions can be used to turn on/off certain actions that the algorithm does. Let's see each one in detail.

Linear Activation Function

In a linear activation, the input data is passed without any change. It is represented by the equation

$$f(x) = x$$

where x is the input given to the activation function. The derivative of this function, $f(x)$, equals 1. The derivative values are useful while performing gradient descent updates in backpropagation. We will discuss backpropagation in a later section. A linear function could not learn a non-linear data type (Figure 5-9). We will discuss non-linear functions, starting with Sigmoid.

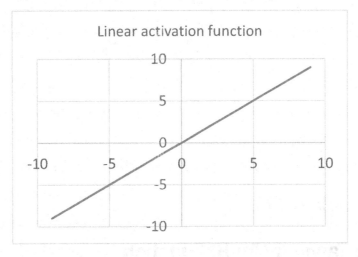

Figure 5-9. *Linear activation function*

Sigmoid Activation Function

In a Sigmoid function, the output is squished between 0 and 1 using the following equation (Figure 5-10):

$$f(x) = \frac{1}{1 + \exp(-x)}$$

We can convert the $f(x)$ value to on/off signal using a customized cut-off, as follows:

$$\begin{cases} 1 \ for \ f(x) \geq cut - off \\ 0 \ for \ f(x) < cut - off \end{cases}$$

When it comes to binary classification, most algorithms use 0.5 as the cut-off value. You can tweak it depending on your exercise. The derivate of this function is

$$f'(x) = f(x)(1 - f(x))$$

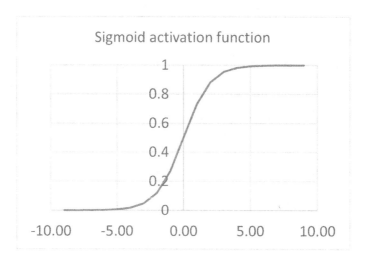

Figure 5-10. *Sigmoid function*

Hyperbolic Tangent (TanH) Function

In a TanH function, the output is squished between -1 and 1 using the following equation (Figure 5-11):

$$f(x) = tanh(x) = \left(\frac{2}{1 + exp(-2x)} \right) - 1$$

Similar to the Sigmoid function, you can use a custom cut-off (default 0) to convert the output to a 0/1 binary output. The derivate of this function is

$$f'(x) = 1 - f(x)^2$$

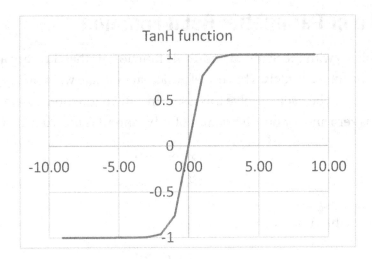

Figure 5-11. *TanH function*

Rectified Linear Unit (ReLu) Function

Rectified ReLu will output the input directly if it is positive, otherwise, it will output zero. The rectified linear activation function overcomes the vanishing gradient problem, allowing models to learn faster and perform better.

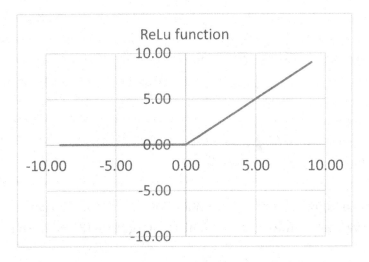

Figure 5-12. *ReLu function*

Leaky ReLu or Parametric ReLu Function

One of the problems with the Sigmoid and TanH functions is that the derivative tends to become 0 for some deep models. The model parameter update will not happen because of this issue. ReLu can be used to solve this issue. In a ReLu function, for negative values of x the output is zero and for positive values of x the output is the same as x.

$$f(x)=\begin{cases} x \ for \ x \geq 0 \\ 0 \ for \ x < 0 \end{cases}$$

The derivate of this function is

$$f'(x)=\begin{cases} 1 \ for \ x \geq 0 \\ 0 \ for \ x < 0 \end{cases}$$

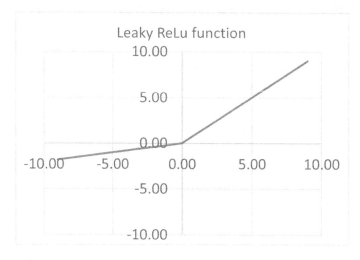

Figure 5-13. *Leaky ReLu*

In a leaky ReLu function, for negative values of x the output is a constant factor of x and for positive values of x the output is same as x (Figure 5-13), where α is the constant factor.

$$f(x)=\begin{cases} x \ for \ x \geq 0 \\ \alpha x \ for \ x < 0 \end{cases}$$

The derivate of this function is

$$f'(x) = \begin{cases} 1 \; for \; x \geq 0 \\ \alpha \; for \; x < 0 \end{cases}$$

Swish Activation Function

This is a relatively new activation function. It was introduced in 2017 by the Google brain team. It is shown in research that by replacing the ReLu function with the Swish function the classification accuracy is improved in the *ImageNet model* by 0.9 percent and in the *Inception_ResNet-v2 model* by 0.6 percent (Figure 5-14). The equation for the function is provided here:

$$f(x) = x * sigmoid(x) = x * \left(\frac{1}{1 + exp(-x)} \right)$$

The derivative of this function is given by

$$f'(x) = f(x) + sigmoid(x) * (1 - f(x))$$

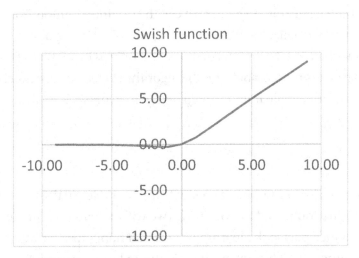

Figure 5-14. *Swish function*

Softmax Function

The equation for this function is

$$f(x) = \frac{e^{Xi}}{\sum_{j} e^{Xj}}$$

When the number of classes is 3, and the x values are [5, 4, 3], $f(x)$ is calculated as

$$f(5) = \frac{e^5}{e^5 + e^4 + e^3} = 0.665$$

For other classes, the values are 0.24 and 0.9, respectively. This activation function is used for multi-label classification problems. The class with the higher probability is chosen as the final output. In this case, the model will pick "Class 1" as output.

Batch Normalization

We've discussed the advantages of using mini-batch in a model-training process. However, it becomes tricky when the distribution of the inputs varies across batches. This will cause the algorithm to train a little bit longer to arrive at the optimal solution. You can resolve this problem using batch normalization. The concept is simple. The input data is standardized for each batch, and the model is trained on the standardized batch data. This will make the model's learning process more stable, and the algorithm will converge much faster. This concept is more powerful while training deep neural networks.

Dropout

Sometimes a model can overemphasize a single input and leave the rest out in the training process. This could lead to overfitting (we will explore this concept in the next chapter) and introduce bias in the learning process. It happens a lot when you are dealing with continuous and binary input variables to train the model. Let's say you have *age* and *gender* as input. In this case, the model would give more weight to age since it has more unique values and would not give similar weight to gender. To overcome this problem, you can use a regularization technique called Dropout. During training, some of the hidden layer (neural network) outputs are randomly dropped. This way, the algorithm will not always rely on the *age* variable to predict the output, thereby avoiding bias and overfitting.

Supervised Machine Learning Algorithms

We have learned all the relevant concepts. Now, let's deep dive into each of the machine learning algorithms. For this entire section, we will use the bank dataset. The code that follows is the starter code for all the machine learning algorithms.

```
# read the data
filename = "bank-full.csv"

from pyspark.sql import SparkSession
spark = SparkSession.builder.getOrCreate()
data = spark.read.csv(filename, header=True, inferSchema=True, sep=';')
data.show()

# assemble feature vectors
from pyspark.ml.feature import VectorAssembler

# assemble individual columns to one column - 'features'
def assemble_vectors(df, features_list, target_variable_name):
    stages = []
    #assemble vectors
    assembler = VectorAssembler(inputCols=features_list,
    outputCol='features')
    stages = [assembler]
    #select all the columns + target + newly created 'features' column
    selectedCols = [target_variable_name, 'features'] + features_list
    #use pipeline to process sequentially
    pipeline = Pipeline(stages=stages)
    #assembler model
    assembleModel = pipeline.fit(df)
    #apply assembler model on data
    df = assembleModel.transform(df).select(selectedCols)
    return df
```

Linear Regression

Linear regression is used to predict continuous outcomes; for example, to predict house price or to predict a customer's lifetime value. This algorithm assumes that the relationship between inputs and outputs is linear. When you use one input to predict the output, it is called simple linear regression. When you use multiple inputs to predict the output, it is called multiple linear regression. The concept is shown in Figure 5-15.

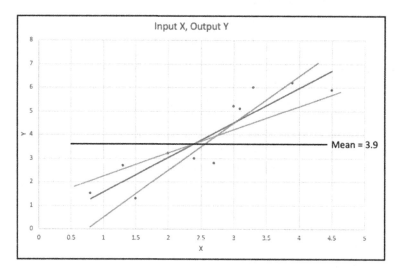

Figure 5-15. *Linear regression*

Looking at the figure, which line better fits the data? Our goal is to get a best possible estimate of *Y* based on *X* values. In this scenario, the blue line fits the data much better than all the other lines provided. The black line is the mean of the *Y* value, which in this case is 3.9. Other lines (orange and grey) are not a good fit. Well, how did I come to this conclusion? I used the squared error algorithm to calculate the error for each of the lines and chose the one that gave me the lowest error. These steps are a simple linear regression model. Since I looked for a line with the least squared error, this is also called an ordinary least squares or least squares regression. By default, PySpark uses *Squared Error* as the loss function. Linear regression can be represented with the following equation:

$$y = \beta_0 + \beta_1 * x_1 + \beta_2 * x_2 + \ldots\ldots\ldots + \beta_n * x_n$$

where, β_0 is the intercept and n is the number of variables in the linear regression model. Let's fit a linear regression model using PySpark and analyze the output. The target is the *balance* variable and the inputs are the rest of the continuous variables: *age, day, duration, campaign, pdays, previous*.

```
#select the variables
linear_df = data.select(['age', 'balance', 'day', 'duration', 'campaign',
'pdays', 'previous'])
target_variable_name = 'balance'

#exclude target variable and select all other feature vectors
features_list = linear_df.columns
#features_list = char_vars #this option is used only for ChiSqselector
features_list.remove(target_variable_name)

# apply the function on our DataFrame
df = assemble_vectors(linear_df, features_list, target_variable_name)

# fit the regression model
from pyspark.ml.regression import LinearRegression
reg = LinearRegression(featuresCol='features', labelCol='balance')
reg_model = reg.fit(df) # fit model
# view the coefficients and intercepts for each variable
import pandas as pd
for k, v in df.schema["features"].metadata["ml_attr"]["attrs"].items():
    features_df = pd.DataFrame(v)
# print coefficient and intercept
print(reg_model.coefficients, reg_model.intercept)
features_df['coefficients'] = reg_model.coefficients

# prediction result
pred_result = reg_model.transform(df)
```

The regression coefficients are provided in Figure 5-16 (features_df dataset).

	idx	name	coefficients
0	0	age	28.083973
1	1	day	3.305546
2	2	duration	0.248828
3	3	campaign	-14.142676
4	4	pdays	-0.082488
5	5	previous	23.462993

Figure 5-16. *Regression coefficients*

To represent this model in an equation, you would write the following:

```
Predicted_balance = 124.92 + 28.08 * age + 3.30 * day +
        0.25 * duration + -14.14 * campaign
        -0.08 * pdays + 23.46 * previous
```

Interpreting the Linear Regression Model

Let's see the significance of the *age* variable in the model. For this purpose, we will take some example values for each input and substitute them into the preceding equation.

{Age:33, day:5, duration:76, campaign:1, pdays:-1, previous:0}

```
Predicted_balance = 124.92 + 28.08 * 33 + 3.30 * 5 +
        0.25 * 76 + -14.14 * 1
        -0.08 * -1 + 23.46 * 0 = 1073
```

That's perfect. Now, let's increase the age value to 34 and keep the rest constant. Doing the same calculation again, we get

```
Predicted_balance = 1101.8
```

```
Difference in prediction due to age increase = 1101.8 - 1073 = 28.08
```

Okay, now we understand the impact of age. To put it in layman's terms, when age increases by 1, the predicted balance increases by 28 (rounded value) and vice-versa. This is easily translatable into business logic, and decisions can be made based on this value. We can interpret other input variables in a similar fashion. The main things to consider while interpreting a linear regression model are the following:

1) Keep all other variables constant except the variable you are trying to interpret.

2) Pick the coefficient of that input variable and use it for interpretation in layman's terms.

Perfect. However, there is a slight problem here, which we will discuss next.

Multi-collinearity

While interpreting a regression model, we keep all the variables constant except one variable. What if the variable that we are trying to interpret is correlated with another input variable in our dataset? Let's assume that age and duration are correlated in our data. In this case, when you increase the *age* variable inherently the *duration* value also increases. You cannot keep the *duration* variable constant in this case. Therefore, the interpretability is lost in the process. This problem is called multi-collinearity.

When two or more variables are correlated, the model results are not interpretable. In a linear regression model, we should avoid multi-collinearity issues. Otherwise, the model output is not valid.

We can check for multi-collinearity using the Variance Inflation Factor (VIF). The industry standard for VIF's value is 10. You can tweak this value as per your needs. When the VIF value is 1, the input features are completely uncorrelated. PCA components usually have a VIF close to 1. In our final model, we should include input variables that end up having a VIF value of less than 10. The formula to calculate VIF is provided here:

$$VIF = \frac{1}{1-R^2}$$

The code for VIF in PySpark is provided here:

```
def vif_calculator(df, features_list):
    vif_list = []
    for i in features_list:
        temp_features_list = features_list.copy()
        temp_features_list.remove(i)
        temp_target = i
        assembler = VectorAssembler(inputCols=temp_features_list,
        outputCol='features')
```

```
        temp_df = assembler.transform(df)
        reg = LinearRegression(featuresCol='features', labelCol=i)
        reg_model = reg.fit(temp_df) # fit model
        temp_vif = 1/(1 - reg_model.summary.r2)
        vif_list.append(temp_vif)
    return vif_list
features_df['vif'] = vif_calculator(linear_df, features_list)
print(features_df)
```

Note   `linear_df` *is provided instead of the* `df`. *Make sure the dataset does not have a column named* `features` *already present. Otherwise, this function will throw an error.*

---

	idx	name	coefficients	vif
0	0	age	28.083973	1.000917
1	1	day	3.305546	1.034350
2	2	duration	0.248828	1.007627
3	3	campaign	-14.142676	1.039907
4	4	pdays	-0.082488	1.276182
5	5	previous	23.462993	1.261321

Figure 5-17. *VIF*

The VIF for all the inputs is less than 10 (Figure 5-17). Therefore, there are no multi-collinearity issues in our model features. When you want to look for the best subset features that produce the best linear regression model, you need to look at Mallows's C_p factor. This concept is out of the scope of this book. In the next chapter, we will look at different ways to validate a regression model and pick the best one.

Logistic Regression

Logistic regression is a classification algorithm. We use this algorithm to predict binary outcomes—whether a person will buy or not, churn versus not churn, spam versus not spam, and so on. Similar to a linear regression, it fits a line on the data to separate the binary outcomes. However, the best-fitting line is decided based on the Maximum Likelihood Estimate. First, let's look at the equation for a logistic regression model:

$$log\left(\frac{p}{1-p}\right) = \beta_0 + \beta_1 * x_1 + \beta_2 * x_2 + \ldots\ldots + \beta_n * x_n$$

The equation for logistic regression differs from that for linear regression on the left-side component. This component, $log\left(\frac{p}{1-p}\right)$, is called *log(odds)*. Why *log(odds)*? Look at Figure 5-18.

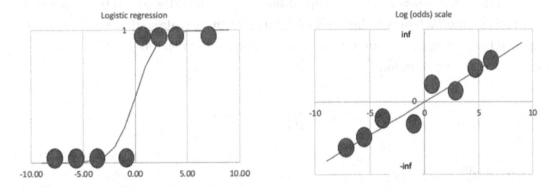

Figure 5-18. *Logistic regression scale versus log odds scale*

Figure 5-18 on the left shows the data that we want to fit the regression line. By converting the 0/1 scale to a *log(odds)* scale, we can map our data in the range (-infinity, + infinity). In this scale, the logistic regression fits the best line.

```
Log(odds) positive range = log(1/0) = log(1) - log(0) = +Infinity
Log(odds) negative range = log(0/1) = log(0) = -Infinity
Log(odds = 0.5) = Log(0.5/0.5) = log(1) = 0
```

Therefore, 0.5 probability is the midpoint. As mentioned, the best fit is determined based on the Maximum Likelihood Estimation. So, what is it?

Maximum Likelihood Estimation or Maximum Log-likelihood (MLE)

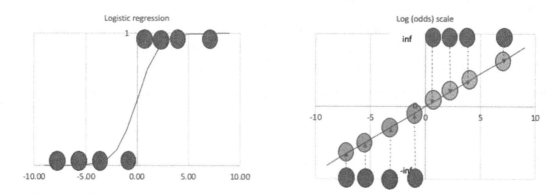

Figure 5-19. *Maximum Likelihood Estimation*

We pick a candidate line as shown on the left of Figure 5-19 and project our data points onto that line. In this process, we end up getting a *log(odds)* value for each data point based on the line. We convert the *log(odds)* to a probability scale so that we can convert the fitted line to a Sigmoid function similar to the one shown on the left. This is done using the following equation:

$$log\left(\frac{p}{1-p}\right) = log(odds)$$

This means

$$\frac{p}{1-p} = e^{log(odds)}$$

Solving this equation, we get

$$p = \frac{e^{log(odds)}}{1+e^{log(odds)}} = \frac{1}{1+e^{-log(odds)}} = \frac{1}{1+e^{-x}} = sigmoid\ function$$

where, $x = log(odds)$. We have the probability values for each observation. Let's start calculating the MLE for this line.

Note   During MLE calculation, use p when the target is 1 and 1- p when the target is 0.

```
Likelihood = prob (1st observation) * prob (2st observation) * ..... prob
(nth observation)
```

where n is the total number of observations in your data and prob is the probability. Taking log on both sides, we get the following:

```
Log(likelihood) = log (prob 1st) + log (prob 2nd) + ...... + log (prob nth)
```

This is the equation we try to maximize in logistic regression. Hence, we call this process Maximum log-likelihood estimation. We repeat this process by rotating the *log(odds)* line and calculating the *log(likelihood)* each time. Finally, the line that maximizes the *log(likelihood)* outcome is the best-fitting line.

How Do You Rotate the Line to Find the Best Fit? Solvers

Solvers can be used to rotate the line in a direction that maximizes the MLE value and reduces the number of iterations required to get a proper fit. In PySpark, the default solver is lbfgs, also known as Limited-memory Broyden Fletcher Goldfarb Shanno.

Binary Versus Multinomial Classification

The Sigmoid function is given by the formula,

$$p(x) = \frac{e^{\log(odds)}}{1 + e^{\log(odds)}} = \frac{e^x}{e^0 + e^x}$$

Here, the e^0 corresponds to the baseline model, which is the 0 class in a binary outcome. e^x corresponds to the 1 class. In a similar way, we can apply the same formula to a multi-class problem. Let's say you have classes a, b, and c. The probability of each class is given by

$$p(a) = \frac{e^{\log(odds)}}{1 + e^{\log(odds)}} = \frac{e^a}{e^a + e^b + e^c}$$

For other classes, the corresponding e element goes to the numerator. This is the SoftMax function, as discussed earlier. Now, you know the link between Sigmoid and SoftMax. Another thing to consider for binary versus multi-class is a link function. In general, we use the logit ($\log\left(\dfrac{p}{1-p}\right)$) link function for binary targets. For multinomial, you can use probit for nominal targets and clog-log for ordinal targets. In PySpark, you can accomplish a binary or multinomial fit by setting the *family* option in the logistic regression algorithm.

PySpark Code

For this exercise, the target variable is y in the bank dataset and the input predictors are *age, balance, day, duration, campaign, pdays, previous*. Target y is a binary class.

```
+---+-----+
|  y|count|
+---+-----+
|0.0|39922|
|1.0| 5289|
+---+-----+
```

```
target_variable_name = "y"
logistic_df = data.select(['age', 'balance', 'day', 'duration', 'campaign',
'pdays', 'previous', 'y'])
#exclude target variable and select all other feature vectors
features_list = logistic_df.columns
#features_list = char_vars #this option is used only for ChiSqselector
features_list.remove(target_variable_name)
# apply the function on our dataframe
df = assemble_vectors(logistic_df, features_list, target_variable_name)
```

We will showcase both binary and multinomial fit using the same target, although it is preferable to use multinomial fit when the target has more than two classes.

```
import numpy as np
from pyspark.ml.classification import LogisticRegression
binary_clf = LogisticRegression(featuresCol='features', labelCol='y',
family='binomial')
```

```
multinomial_clf = LogisticRegression(featuresCol='features', labelCol='y',
family='multinomial')
binary_clf_model = binary_clf.fit(df) # fit binary model
multinomial_clf_model = multinomial_clf.fit(df) # fit multinomial model
np.set_printoptions(precision=3, suppress=True)
#model coefficients for binary model
print(binary_clf_model.coefficients)
#model coefficients for multinomial model
np.set_printoptions(precision=4, suppress=True)
print(multinomial_clf_model.coefficientMatrix)
print(binary_clf_model.intercept) #model intercept for binary model
#model intercept for multinomial model
print(multinomial_clf_model.interceptVector)
```

The equation for the binary model is provided here:

log(odds class1) = -3.47 + 0.008*age + 0*balance -0.0017*day +
0.0036*duration - 0.128*campaign + 0.0021*pdays + 0.0859*previous

Similarly, the equation for the multinomial model is provided here:

log(odds class0) = 1.735 -0.004*age – 0*balance + 0.0008*day -
0.0018*duration + 0.064*campaign - 0.0011*pdays - 0.043*previous

and

log(odds class1) = -1.735 +0.004*age + 0*balance - 0.0008*day +
0.0018*duration - 0.064*campaign + 0.0011*pdays + 0.043*previous

In the multinomial model output, when you subtract the class 1 coefficient from the class 0 coefficient, you get the binary model coefficient for class 1. The same thing applies to intercept too. Okay, now let's interpret the model results

Interpreting the Model Results

Let's go through the same process we used for the linear regression coefficient interpretation, keeping all variables constant and changing the variable we are trying to interpret.

Binary Variable Interpretation

When it comes to binary variables, we calculate the *log(odds)* for both classes using the preceding equation. Let's assume we had gender (Male = 0, Female = 1) in our model. Then, the impact of gender is measured by the following:

```
log(odds class 1|Male) = Intercept + 0.16 * 0 + coef2 * var2 + ......
log(odds class 1|Female) = Intercept + 0.16 * 1 + coef2 * var2 + ......
log(odds class 1|Female) - log(odds class 1|Male) = 0.16
```

By keeping all the other factors constant, we get the coefficient of the binary variable as the difference in *log(odds)*. The best way to interpret the coefficient is using odds. Let's do that.

$$\text{Log}(\frac{\text{odds class 1 Female}}{\text{odds class 1 Male}}) = 0.16$$

$$\frac{\text{odds class 1 Female}}{\text{odds class 1 Male}} = e^{0.16} = 1.173$$

We take the difference between the preceding value and 1 to interpret the odds. The odds of a female being in class 1 is 17 percent higher than the odds of a male being in class 1.

Continuous Variable Interpretation

Let's perform the same calculation with age.

```
log(odds class 1|Age = 29) = Intercept - 0.32 * 29 + coef2 * var2 + ......
log(odds class 1|Age = 30) = Intercept - 0.32 * 30 + coef2 * var2 + ......
log(odds class 1| Age = 30) - log(odds class 1| Age = 29) = -0.32
```

$$\text{Log}(\frac{\text{odds class 1 Age} = 30}{\text{odds class 1 Age} = 29}) = -0.32$$

$$\frac{\text{odds class 1 Age} = 30}{\text{odds class 1 Age} = 29} = e^{-0.32} = 0.73$$

Now, $1 - 0.73 = .27$. When age increases by one unit, the odds of a person's being in class 1 decreases by 27 percent.

Here is the rule of thumb for interpreting odds ratio:

1) Odds ratio of 1 means there is no difference in impact.

2) Odds ratio greater than 1 means the odds increase as the variable changes.

3) Odds ratio less than 1 means the odds decrease as the variable changes.

Decision Trees

So far, we have discussed linear methods for regression and classification. Starting with this section, we will discuss non-linear methods that can be used to solve a problem. Overall, the idea behind decision trees is to generate a set of "if-then rules" and represent it in the form of a tree. Let's visualize a simple decision tree for credit card approval.

Interpretation of the Decision Tree

- The root/parent node of the tree is age variable. Therefore, it is the most important feature for this decision tree.

- The leaf nodes are the decisions that can be taken based on the input variables.

Let's traverse the tree for these conditions (Figure 5-20).

Age = 20, occupation = Student, Credit score = 800	Age = 35, occupation = Employed, Credit score = 750
Age < 25; True Student? True	Age < 25; False Age < 60; True Credit Score > 720
Final verdict = Not approved	Final verdict = Approved

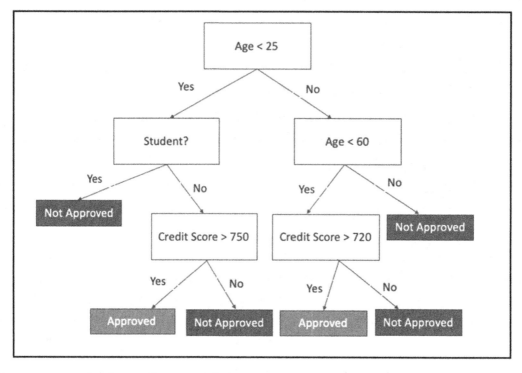

Figure 5-20. *Credit card approval decision tree*

Now, we know that age is the most important attribute. How did the decision tree pick this variable? In order to understand why, we need to cover Gini, entropy, and information gain. Let's go through each one of them (Table 5-3).

Table 5-3. *Algorithm Specifications*

Algorithm/Split criterion	Type of Tree
Gini impurity	CART (Classification and Regression Trees)
Entropy & Information Gain	ID3/C4.5

For this entire exercise, we will use the dataset shown in Table 5-4. The target for our data is a credit card approval with two levels—approved/not approved.

Table 5-4. *Dataset for Exercise*

Age	Gender	Approved/Not approved
30	Male	Not approved
25	Female	Approved
45	Male	Approved
57	Female	Not approved
27	Male	Approved
54	Female	Approved
35	Female	Not approved

Entropy

Entropy is the measure of randomness in the data. The formula to calculate entropy is

$$Entropy = -\sum p(x) \log_2 p(x)$$

It ranges between 0 and 1. Entropy is 0 when there is no randomness and entropy is 1 when the dataset is completely random. Let's use the following example to understand entropy.

Example: We have two bags. In the first bag, there are four blue balls and no red balls. In the second bag, we have two blue balls and two red balls. Calculate the entropy in both the bags.

Bag 1 entropy

```
Probability of picking blue ball in bag 1 = 4/4 = 1
Probability of picking red ball in bag 1 = 0/4 = 0
Entropy = - (1 * log(1) + 0* log(0)) = 0
```

Bag 2 entropy

```
Probability of picking blue ball in bag 2 = 2/4 = 0.5
Probability of picking red ball in bag 2 = 2/4 = 0.5
Entropy = - (0.5 * log(0.5) + 0.5* log(0.5)) = 1
```

Let's extend this example and give you a choice to pick a bag. You win the game when you pick a blue ball. Which bag would you choose?

The answer is obvious now right. You pick the first bag because it has all blue balls. When the entropy is 0, you are certain of the outcome, whereas when the entropy is 1, the outcome is random. We will use this methodology in machine learning to build the decision tree.

Let's switch back to our example data and calculate the entropy for age and gender. This time we will calculate the probability of being approved. For the purpose of this demonstration, let's split the *age* variable based on above 30 and below 30.

Entropy of target (Approved/not approved) – Parent entropy:

```
Probability of approved = 4/7 = 0.57
Probability of not approved = 3/7 = 0.43
Entropy of parent node = - (0.57 * log(0.57) + 0.43 * log(0.43)) = 0.99
```

Entropy of Age Variable:

```
Probability of approved when age >= 30 = 2/5 = 0.4
Probability of not approved when age >= 30 = 3/5 = 0.6
Probability of approved when age < 20 = 2/2 = 1
Probability of not approved when age < 20 = 0/2 = 0

Entropy(Age >= 30) = - (0.4 * log(0.4) + 0.6* log(0.6)) = 0.97
Entropy(Age < 30) = - (1 * log(1) + 0 * log(0)) = 0
```

Entropy of Gender Variable:

```
Probability of approved when female = 2/4 = 0.5
Probability of not approved when female = 2/4 = 0.5
Probability of approved when male = 2/3 = 0.67
Probability of not approved when male = 1/3 = 0.33

Entropy(Gender = Female) = - (0.5 * log(0.5) + 0.5* log(0.5)) = 1
Entropy(Gender = Male) = - (0.67 * log(0.67) + 0.33 * log(0.33)) = 0.92
```

Information Gain

Up until now, we have calculated the entropy for each variable. Now, it's time to decide the variable that goes to the root/parent node. To make this decision, we will use information gain. Information gain tells us the knowledge gained in each split. We are trying to maximize our knowledge on credit card approval. Therefore, we will try to maximize the information gain in each split. It is calculated using the following formula:

```
Information Gain, IG = Entropy (parent node) - (Entropy each child node *
proportion of observations in each child node)
```

Let's calculate the information of each variable. Since five observations went in age >= 30 and two observations in age < 30, the information gain is given by

```
IG (age >= 30)  = 0.99 - (0.97 * 5/7 + 0 * 2/7) = 0.29
```

In a similar way, the IG for gender is given by

```
IG gender = 0.99 - (1 * 4/7 + 0.92 * 3/7) = 0.02
```

The variable *age* has higher information gain than variable *gender*. Therefore, we will pick *age* as our root node. We will start building our tree based on this condition (Figure 5-21).

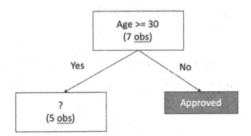

Figure 5-21. *Information gain first split*

Based on our current model, when a person's age is less than 30, the credit card is approved. Our job now is to figure out the next split. We will iterate through the same process again. For *age*, we will use 45 as split point this time.

```
Entropy (Age >=45) = - (0.67 * log(0.67) + 0.33 * log(0.33)) = 0.92
Entropy (Age < 45) = - (0 * log(0) + 1* log(1)) = 0
Entropy (Gender = Female) = -(0.33*log(0.33) + 0.67 * log(0.67))= 0.92
```

```
Entropy (Gender = Male) = - (0.5 * log(0.5) + 0.5* log(0.5)) = 1
IG (age >= 45)  = 0.97 - (0.92 * 3/5 + 0 * 2/5) = 0.418
IG (Gender) = 0.97 - (0.92 * 3/5 + 1 * 2/5) = 0.018
```

Again, information gain for *age* is higher than that for the *gender* variable. So, our next split would look like Figure 5-22.

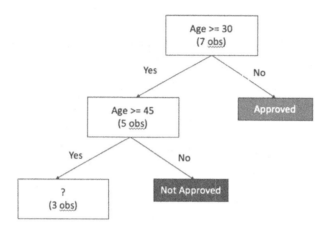

Figure 5-22. *Information gain second split*

For the next split, we will use *age* as 60 and repeat the same process.

```
IG (age >= 60) = 0.92 - (0 * 0 + 0.92) = 0
IG (Gender) = 0.92 - (0 * 1/3 + 1 * 2/3) = 0.25
```

After this split, there are only two observations, and since both are female we cannot use gender anymore. Therefore, we will split based on age, and we get the final tree as shown in Figure 5-23.

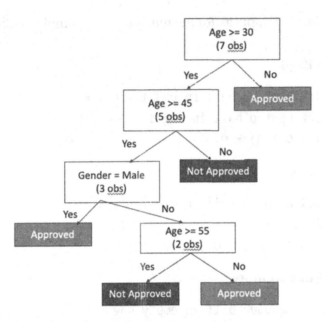

Figure 5-23. *Final tree based on information gain*

That is all. When you get new data, you run through the decision tree to get the output. ID3, C4.5, and C5.0 decision trees are built using entropy/information gain.

Gini Impurity

Let's look at another approach to forming a decision tree, Gini impurity. You should not confuse this concept with the Gini coefficient. Gini impurity measures the impurity in a split. The formula for Gini impurity is as follows:

$$Gini = 1 - \Sigma (p(x))^2$$

The range of the Gini impurity is between 0 and 1.

- When 0, the split is pure and has no impurities.

- When 0.5, the split is random and completely impure. Both classes are equally present in the data.

- Any value above 0.5 can happen when one of the child nodes does not have a single data point in it. In our example dataset, when we consider Age >=60, the Gini impurity for this split will be more than 0.5 since we don't have any data point with age greater than 60.

181

Let's calculate the Gini impurity for the red/blue ball example to understand this concept.

Gini impurity – Bag 1

```
Probability of picking blue ball in bag 1 = 4/4 = 1
Probability of picking red ball in bag 1 = 0/4 = 0
Gini = 1 - (1 * 1 + 0 * 0) = 0
```

Gini impurity – Bag 2

```
Probability of picking blue ball in bag 2 = 2/4 = 0.5
Probability of picking red ball in bag 2 = 2/4 = 0.5
Gini = 1 - (0.5 * 0.5 + 0.5 * 0.5) = 0.5
```

Gini impurity – Bag is empty

```
Probability of picking blue ball in empty bag = 0
Probability of picking red ball in empty bag = 0
Gini = 1 - 0 = 1
```

Again, when you need to pick a bag, you will pick the bag that has a Gini impurity of 0 because you are certain of the outcome. Let's apply the Gini concept to our data to build a decision tree. We will follow the same steps that were used for entropy and calculate the impurity at each split.

```
Gini (Age >= 30) = 1 - (2/5 * 2/5 + 3/5 * 3/5) = 0.48
Gini (Age < 30) = 1 - (2/2 * 2/2 + 0/2 * 0/2) = 0
Weighted Gini for Age > = 30 = (5/7 * 0.48 + 2/7 * 0) = 0.34

Gini (Gender = Female) = 1 - (2/4 * 2/4 + 2/4 * 2/4) = 0.5
Gini (Gender = Male) = 1 = (1/3 * 1/3 + 2/3 * 2/3) = 0.44
Weighted Gini for Gender = (4/7 * 0.5 + 3/7 * 0.44) = 0.48
```

The *age* variable is used for the first split in the tree. Let's continue this process and calculate the final tree. You will get the exact same tree as in entropy and information gain. This is because of the choice of our bins during the age split. However, this is not always the case. In real-world examples, the Gini impurity tree will be different from the entropy and information gain tree. Classification and Regression Trees (CART) trees are built using Gini impurity.

Variance

One more method that PySpark uses to create decision trees is by using variance. The formula for variance is

$$Variance = \frac{\sum\left(X - mean\left(X\right)\right)^2}{n}$$

This concept is used in building regression trees. You can continue splitting the tree by choosing the variable that minimizes the variance at each split. Decision trees can also be made using other criteria like chi-square and correlation. The concept is still the same. We pick variables that maximize the chi-square value in a chi-square. For a correlation tree, we pick variables that minimize the correlation. These concepts are out of the scope of this book.

Determine the Cut-off for Numerical Variables

We used a custom cut-off for age while building the decision tree by hand. In a decision tree algorithm, numeric variables are handled in a different way. For small datasets, each unique value is used as a cut-off point. For large datasets, the numeric variables are binned into a specific number of quantiles. This is set using the maxBins option in the decision tree algorithm. After creating the quantiles, the average value of each bin is used as a cut-off point. Let's assume we choose three bins for the *age* variable and our cut-off points after binning are 25, 42, and 55. The information gain or Gini impurity is calculated at each of these cut-off points, and the point that provides the best outcome is chosen for the split. The same process is done for all numeric variables for each split.

Pruning the Tree/Stopping Criterion

Pruning a tree refers to chopping off the unwanted leaf nodes. This is done to keep the model simple and to avoid overfitting. We will discuss overfitting in detail in the next chapter. Decision trees are prone to overfitting. Unless controlled, you will end up having a tree that memorizes training data. Pruning can be accomplished in a tree using the following options:

- maxDepth – Depth of a tree refers to the number of tree traversal steps required from a root node to reach the leaf node. By default, the maximum depth of a tree is 5.

- minInstancesPerNode – The minimum number of samples to be present in each child node is specified using this option. When you set this option to 5 and a split produces a child node with less than 5 observations, then the split will not happen.

- minInfoGain – A tree split should provide information gain of at least this number specified. When the information gain is less than this number, the split will not happen.

- L1/L2 Regularization – You can use this option to perform pruning of decision trees. We will discuss this in later sections.

PySpark Code for Decision Tree Classification

The dataset used in both the linear and the logistic regression fits is renamed to continuous_df and binary_df, respectively.

```
from pyspark.ml.classification import DecisionTreeClassifier

clf = DecisionTreeClassifier(featuresCol='features', labelCol='y',
impurity='gini') #gini based model
clf_model = clf.fit(binary_df)
clf2 = DecisionTreeClassifier(featuresCol='features', labelCol='y',
impurity='entropy') #entropy based model
clf_model2 = clf2.fit(binary_df)
clf_model.transform(binary_df) #future predictions
# gini feature importance
print(clf_model.featureImportances)

#output - (7,[0,3,5],[0.063,0.723,0.214])
print(clf_model2.featureImportances)
#output - (7,[0,2,3,4,5],[0.018,0.001,0.727,0.0004,0.254])
```

As we can see from the preceding output, the Gini and entropy methods produce different trees and with different sets of variables.

PySpark Code for Decision Tree Regression

To fit a regression tree, you need to set the impurity as a variance. By default, decision tree regression uses the variance method. Unfortunately, PySpark does not support this.

```
from pyspark.ml.regression import DecisionTreeRegressor

reg = DecisionTreeRegressor(featuresCol='features', labelCol='balance',
impurity='variance')
reg_model = reg.fit(continuous_df)
print(reg_model.featureImportances) #feature importance
reg_model.transform(continuous_df) #future predictions
```

Similar to classification trees, we can interpret the feature importance of the input variables. To look at the if-then rules of the decision tree, you can use the following code:

```
clf_model.toDebugString
reg_mode.toDebugString
```

You can plot these decision tree rules using the `Graphviz`, `pydot`, or `networkx` packages in Python.

Feature Importance Using Decision Trees

In the previous chapter, we used decision trees to calculate feature importance. In this section, we will go through the details of how this is calculated.

$$f_i = \sum_{j:nodes\ splits\ on\ feature\ i} s_j * G_j$$

Where

$$f_i = \text{importance of feature}$$

$$s_j = \text{number of samples in node } j$$

$$G_j = \text{impurity value of node } j$$

185

At each node, we calculate the Gini impurity and multiply it by the number of samples that went through the node. If a variable appears more than once in a decision tree, the outputs of each node are added together to get the final feature importance for that variable. Finally, the feature importance of each variable is normalized so that they all add to 1. It is done using the following equation. Let's say we have *n features;* the final feature importance is given by

$$\text{final} - f_i = \frac{f_i}{f_1 + f_2 + \ldots\ldots + f_n}$$

Where

$f_1, f_2, \ldots\ldots, f_n$ are the individual feature importance of each variable

f_i – feature imporatance of the ith variable calculated using the previous equation

$\text{final} - f_i$ – normalized feature importance of variable i

You need to be aware that the feature importance is calculated based on the training dataset. Therefore, the feature importance could be biased when the model is overfitting. To overcome this, you can use proper validation techniques or use permutation importance to calculate the feature importance. We will talk about validation techniques in the next chapter. The permutation importance concept is out of the scope of this book. We can now move on to other tree-based techniques.

Random Forests

In a decision tree, we build a single tree based on the training data. Instead of a single tree, what if we build multiple trees and aggregate the results of each tree to get the final prediction? Well, that is the idea behind random forests. A random forest is an ensemble model built on a concept called bagging.

- *Ensemble* means a collection of different models. The individual models can be built using a variety of machine learning algorithms like logistic regression, decision tree, neural networks, and so on. The final result is determined based on voting (in case of classification) or the average of predictions (in case of regression).

- Bagging, a.k.a bootstrap aggregation, is a concept by which random subsets of data (both rows and columns) are used to train a model. In a random forest algorithm, multiple decision trees are trained on different subsets of the same dataset. For example, the first tree might train using age and gender as input variables, the second tree might train using balance and income as input variables, and so on. In addition, the rows (samples) used to train the first and second tree are randomly sampled from the original training dataset. You can sample rows *with or without replacement.* When you choose to go *with replacement,* the same sample can be used more than once. In a *without replacement* scenario, the same sample cannot be used more than once.

Random forests use both these concepts to build trees. The process is as follows:

1. Select random features – subset features (featureSubsetStrategy)

2. Select random samples (with or without replacement) – subset rows (subsamplingRate)

3. Form the subset data based on the subset features and rows.

4. Build a decision tree based on the subset data.

5. Repeat the process to build the number of trees specified by the user. The PySpark default for number of trees is 20. (numTrees)

6. For final prediction using random forest, run the data through all the individual trees. Get the prediction of each tree.

7. For classification, count the number of votes for each class. The class with the maximum votes is chosen as final output.

8. For regression, average the output of individual trees to get the final output.

Hyperparameter Tuning

By tuning the featureSubsetStrategy and subsamplingRate parameters, the model performance and training process speed can be improved.

featureSubsetStrategy - "auto", "all", "sqrt", "log2", "onethird"

subsamplingRate – any value between 0 and 1. When the value is set to 1, the whole dataset is used.

Feature Importance Using Random Forest

In random forests, feature importance is calculated in a similar way as for a decision tree, with one slight change. The average of the feature importance of a variable is calculated before the normalization step.

$$average - f_i = \frac{f_i}{\text{number of trees with feature } f_i}$$

The *average – f_i* for all variables is used to calculate the final feature importance, *final – f_i*.

The same concept is used in gradient-boosting trees as well, which we will see next.

PySpark Code for Random Forest

Classification

```
from pyspark.ml.classification import RandomForestClassifier
clf = RandomForestClassifier(featuresCol='features', labelCol='y')
clf_model = clf.fit(binary_df)
print(clf_model.featureImportances)
print(clf_model.toDebugString)
```

Regression

```
from pyspark.ml.regression import RandomForestRegressor
reg = RandomForestRegressor(featuresCol='features', labelCol='balance')
reg_model = reg.fit(continuous_df)
print(reg_model.featureImportances)
print(reg_model.toDebugString)
```

Why Random Forest?

So far, we have not covered why we need random forests. Let's go through it now.

- Random forests are more robust than a single decision tree and they limit overfitting.

- They remove feature selection bias by randomly selecting features during each tree training process.

- Random forest uses a proximity matrix, which can be used to fill in missing values.

Gradient Boosting

Another variant of the decision tree is gradient-boosted trees. Gradient boosting is an ensemble model that is built on the concept called boosting. We already went through the ensemble concept. Let's understand boosting in more detail.

- Boosting uses weak learners to build trees. It is an additive modeling technique. It is also referred to as sequential learners because the current tree learns from the mistakes of the previous tree.

Boosting Learning Process

- Model training starts by building an initial decision tree. Predictions are made using this tree.

- A sample weightage column is created based on the prediction output. When there is an error in prediction, the sample is given more weightage, and when the predictions are accurate, the sample is given less weightage.

- We create a new training dataset based on the sample weightage column. This is to ensure that the error samples are given more preference in the newly created training data.

- We repeat the training process with this new data and continue building trees until the user-specified numTrees option is met.

- Learning rate (learningRate) is used to scale each weak learner output.

- Final model prediction is given by the following:

```
First tree prediction + learningRate*Second tree prediction + ....... +
learningRate * Nth tree prediction
```

PySpark Code for Gradient Boosting

Classification

```
from pyspark.ml.classification import GBTClassifier
clf = GBTClassifier(featuresCol='features', labelCol='y')
clf_model = clf.fit(binary_df)
print(clf_model.featureImportances)
print(clf_model.toDebugString)
```

Regression

```
from pyspark.ml.regression import GBTRegressor
reg = GBTRegressor(featuresCol='features', labelCol='balance')
reg_model = reg.fit(continuous_df)
print(reg_model.featureImportances)
print(reg_model.toDebugString)
```

Why Gradient Boosting?

- Useful for modeling imbalanced target classes

- Gradient boosting builds shallow trees, as compared to the deep trees created by random forest and decision trees. Therefore, it is good for reducing bias in predictions.

Support Vector Machine (SVM)

Imagine you are in a classroom full of students. Your task is to draw a line that separates boys from girls. In addition to drawing the line, your method should also have these characteristics:

- A good margin of separation between boys and girls

- A low error rate during separation

This is accomplished using support vector machines (SVM). In general, support vector machines use hyperplanes to perform classification. Let's look at Figure 5-24 to understand the concept in detail.

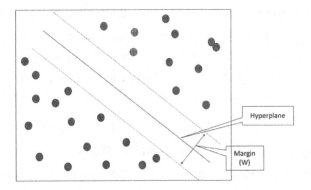

Figure 5-24. *Feature space separation using a support vector machine*

In the figure, a linear hyperplane is used to separate the classes. For non-linear data, we can use polynomial or rbf (radial basis function) hyperplanes to get a good separation. This is called a kernel trick. Currently, PySpark supports only linear kernel. However, Python version supports other kernels. The SVM model comprises two error functions, as follows (Figure 5-25):

```
Total error = Classification error + Margin error
```

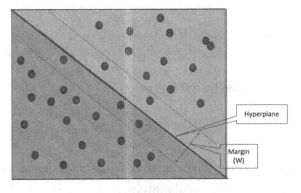

Figure 5-25. *Classification and margin errors*

Classification Error

Classification error measures the error in prediction in the SVM model. This error is similar to misclassification rate. However, we mentioned that the margin is also a component of SVM. Therefore, any observation that falls within the margin is also considered an error since we are trying to maximize the margin. In Figure 5-25, there are seven samples that contribute to classification error: two outside the margin and five within the margin. The total classification error is the sum of the absolute value of the distances of these observations from the margin boundaries. In order to calculate the red observation error, we will calculate distance from the upper margin, and for the blue observation error, we will calculate distance from the lower margin.

Margin Error

Margin error is going to quantify the error associated with the margin boundaries. When you have a large margin, the error is small and vice-versa. Let's represent the line in the form of an equation, as follows:

$$ax + by + c = 0 --- > center\ line$$

$$ax + by + c = i --- > upper\ margin$$

$$ax + by + c = -i --- > lower\ margin$$

We can calculate the width of the margin using the following formula:

$$W = \frac{2i}{\sqrt{a^2 + b^2}}$$

And the margin error is given by the following:

$$Margin\ error = a^2 + b^2$$

When you substitute values in the preceding equation, you get a small error for a large margin and a large error for a small margin.

Total Error

As we saw before, total error is the sum of the classification and margin errors. In general, a large margin produces a large classification error and vice-versa. You can tweak the model error function by using a regularization parameter (regParam). We will focus on regularization in later sections.

PySpark Code

```
from pyspark.ml.classification import LinearSVC
clf = LinearSVC(featuresCol='features', labelCol='y')
clf_model = clf.fit(binary_df)
print(clf_model.intercept, clf_model.coefficients)
```

Note   SVM takes a long time to train. You can play with the regParam option to train the model faster.

Neural Networks

Neural networks are inspired by the biological neurons in the human brain. In humans, brains communicate with other parts of the body using these neurons. Depending upon the task, the brain activates the neurons corresponding to the task and accomplishes it. Artificial neural networks (ANNs) are built on this concept and use activation functions to activate certain neurons and get the required output. A simple neural network architecture is shown in Figure 5-26.

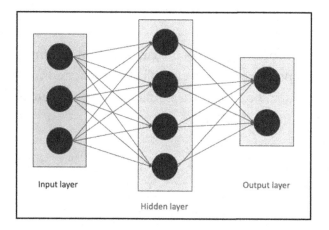

Figure 5-26. *ANN architecture*

Things to Know About ANN Architecture

- The simplest neural networks include input and output layers without a hidden layer. This type of network is called a perceptron (Figure 5-27). A linear or logistic regression model can be built with a perceptron neural network.

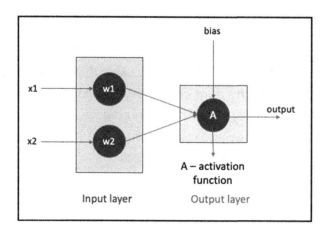

Figure 5-27. *Perceptron*

- When the activation function is linear, the perceptron produces a linear regression, and when the activation function is Sigmoid, the perceptron produces a logistic regression.

- We can add multiple hidden layers between input and output layers in a perceptron. Because of this capability, neural networks are also called multi-layer perceptrons.

- A single unit in a hidden layer is called a hidden unit. In Figure 5-26, there are four hidden units in the hidden layer.

- You can use activation functions to turn on/off a hidden unit. When you use a linear activation function, it models a linear relationship. By using non-linear activation functions like ReLu and TanH, you can model a complex relationship.

- When you stack multiple hidden layers, you produce a deep neural network.

- When you increase the size of the hidden units in a hidden layer, you produce a wider neural network.

- Neural networks do not need prior knowledge about the data. They can model the relationship between features and targets using these two steps: feed forward (step 1), backpropagation and gradient descent (step 2). This is the reason neural networks are also called feed forward neural networks.

- In PySpark, neural networks can be optimized using the hyperparameters `maxIter`, `layers`, `blockSize`, `stepSize`

How Does a Neural Network Fit the Data?

Let's use logic gates to understand the concept in detail. We will use an OR gate for our example. For this exercise, we will use a perceptron network. Table 5-5 contains the data.

Table 5-5. *Sample Data*

Input 1	Input 2	Output
0	0	0
0	1	1
1	0	1
1	1	1

After a neural network architecture is set up, the bias and weights are randomly initialized. In general, a gaussian or uniform function is used to initialize bias and weights. We will set the bias as 1 for our example. Since we have two features in our example, we will use two weights (w1 and w2). Let's assume the initial random weight values are w1 – 0.1 and w2 – 0.7. The activation function for this example is Sigmoid.

$$\text{Sigmoid function}, \sigma = \frac{1}{1 + \exp(-x)}$$

$$\textit{final output} = \begin{cases} 0 \ \textit{if} \ \sigma < 0.5 \\ 1 \ \textit{if} \ \sigma \geq 0.5 \end{cases}$$

Neural network model output $= w1 * x1 + w2 * x2 + b$

Step 1: Feed Forward

In a feed forward step, the weights are multiplied by the inputs and sent to the output function to calculate the final output (Table 5-6).

Table 5-6. *Error Calculations*

Input1	Input2	Target	Output = 0.1 * Input1 + 0.7 * Input2 + 1	Sigmoid(output) = $\frac{1}{1 + exp(-output)}$	Error = Target - Sigmoid(output)
0	0	0	1	0.73	-0.73
0	1	1	1.7	0.85	0.15
1	0	1	1.1	0.75	0.25
1	1	1	1.8	0.86	0.14

Step 2: Backpropagation

Let's backpropagate the errors to update the weights of the input features. To update the weights, we use the following formula:

```
New weight = old weight - learningrate * error delta
```

```
Error delta = input * error * derivative(output)
```

We will set the learning rate as 1. As we see from the preceding equation, we take the derivate of each error output to update the weight (Table 5-7). The derivative of the Sigmoid function is given by the following:

$$\text{Derivation of Sigmoid function}, \sigma' = \sigma * (1 - \sigma)$$

Table 5-7. *Derivative Calculations*

Error	I = Derivative(sigmoid)	J = Error * I	W1_Delta = Input1 * J	W2_Delta = Input2 * J
-0.73	0.19	-0.14	0	0
0.15	0.13	0.02	0	0.02
0.25	0.18	0.05	0.05	0
0.14	0.12	0.02	0.02	0.02

```
W1_errorDelta = sum(W1_Delta) = 0.07
W2_errorDelta = sum(W2_Delta) = 0.04
New_w1 = 0.1 - 0.07 = 0.03
New_w2 = 0.7 - 0.04 = 0.66
```

That's it. We have our first iteration (epoch) complete, and we have the updated weights after one epoch. We repeat the steps for a certain number of epochs to train our model. During each step, the weight gets updated, and slowly the model learns to fit according to the data.

Hyperparameters Tuning in Neural Networks

- `maxIter` – Number of epochs to run. One epoch includes both feed forward and backpropagation. When the number of epochs is more, the model takes more time to train, and vice-versa. In addition, when the number of epochs is very low, the model will not be able to learn the patterns in the data. On the other hand, by choosing a larger epoch the model is prone to overfitting.

- `layers` – Number of hidden layers to use while training the model.

- `Format` – [hidden layer1, ..., Hidden layer *n*, output layer]. When the hidden layer increases, we can get more-accurate predictions. However, the model is prone to overfit the data. When the layer decreases, the model won't have enough space to learn. The trick is to find the right balance. A good number to start would be *n* or 2*n*, where *n* is the number of input features.

- `blockSize` – Batch size. The number of batches to be used while training the model. This is the mini-batch concept we discussed before.

- `stepSize` – Learning rate for gradient descent. How fast should your model learn? We already discussed this in detail in the previous section.

PySpark Code

```
from pyspark.ml.classification import MultilayerPerceptronClassifier
#output_layer is set to 2 because of binary target
clf = MultilayerPerceptronClassifier(featuresCol='features', labelCol='y',
layers=[4, 4, 2])
clf_model = clf.fit(binary_df)
```

Currently, PySpark does not support regression using neural networks. However, you can implement regression using a linear activation function.

One-vs-Rest Classifier

This is used for multi-class classification exercises. The idea behind this classifier is very simple. Let's say you have four classes for *education—primary, secondary, tertiary, and unknown*. Your objective is to predict the *education* outcome for an individual. This is a multi-class problem because you have more than two output classes. We create a base classifier and use that base classifier to perform binary classification for each class. The first classifier will predict primary versus non-primary (secondary, tertiary, unknown). The second classifier will predict secondary versus non-secondary (primary, tertiary, unknown). In a similar fashion, the third classifier will predict tertiary versus non-tertiary, and the fourth classifier will predict unknown versus non-unknown. The final prediction is obtained by evaluating these four individual classifiers and choosing the index of the most confident classifier as output.

PySpark Code

```
target_variable_name = "education"
multiclass_df = data.select(['age', 'balance', 'day', 'duration',
'campaign', 'pdays', 'previous', 'job', 'education'])
features_list = multiclass_df.columns
#features_list = char_vars #this option is used only for ChiSqselector
features_list.remove(target_variable_name)
# apply the function on our dataframe
multiclass_df = assemble_vectors(multiclass_df, features_list, target_
variable_name)

# fitting the one-vs-rest classifier
from pyspark.ml.classification import RandomForestClassifier, OneVsRest
from pyspark.ml.evaluation import MulticlassClassificationEvaluator
# generate the train/test split.
(train, test) = multiclass_df.randomSplit([0.7, 0.3])
# instantiate the base classifier.
clf = RandomForestClassifier(featuresCol='features', labelCol='education')
# instantiate the One Vs Rest Classifier.
ovr = OneVsRest(classifier=clf, featuresCol='features',
labelCol='education')
# train the multiclass model.
ovrModel = ovr.fit(train)
# score the model on test data.
predictions = ovrModel.transform(test)
# obtain evaluator.
evaluator = MulticlassClassificationEvaluator(metricName="accuracy",
labelCol='education')
# compute the classification error on test data.
accuracy = evaluator.evaluate(predictions)
print("Test Error = %g" % (1.0 - accuracy))
```

Naïve Bayes Classifier

The Naïve Bayes classifier is based on Bayes theorem. It assumes the independence of the input variables. This is the reason it is called *naïve*—it is not always true with the input variables. Bayes' theorem is used to calculate posterior probability based on prior probability and likelihood. It is denoted by the following:

$$P(A|B) = \frac{P(B|A) * P(A)}{P(B)}$$

Where,

A & B are the events.

P(A|B) = posterior probability. It is read as "probability of A given B is true

P(B|A) = Likelihood. It is read as "probability of B given A is true"

P(A) = Prior probability of event A

P(B) = Prior probability of event B

Let's look at the data in Table 5-8 to understand the Bayes theorem. We have two buckets. Each of them has ten balls. The first bucket has seven red balls and three green balls. The second bucket has seven green balls and three red balls. Let's assume we have an equal chance of choosing the buckets. Assume you randomly took a ball from one of the buckets and it came out as a green ball. What is the probability that the ball came from bucket 2?

Table 5-8. *Bayes Demonstration—Sample Data*

Bucket1	Bucket2
Red	Red
Red	Green
Red	Green
Green	Green
Red	Green
Red	Green
Green	Red
Red	Red
Red	Green
Green	Green

There are multiple ways to find the answer for this question. I will use the Bayes theorem for this problem since it is most suitable for these cases.

- We have two events: A – *bucket2* and B – green color ball.

- We have equal chances of selecting buckets, so prior probability of *bucket 2* = P(A) = 0.5

- Prior probability of choosing green color ball (GB) which is event B is given by P(B)

- P(B) = P(*bucket1*) * P(GB in *bucket1*) + P(*bucket2*) * P(GB in *bucket2*)

- P(B) = 0.5 * 0.3 + 0.5 * 0.7 = 0.5

- The likelihood of a green ball given *bucket2* is given by

$$P(Greenball \,|\, Bucket2) = P(B|A) = 0.7$$

- Probability of *bucket2* given green ball:

$$P(Bucket2|Greenball) = P(A|B) = \frac{P(B|A)*P(A)}{P(B)} = (0.7 * 0.5)/0.5 = 0.7$$

Bayes' theorem provides a simple solution for these types of problems. The classifier is an extension of this theorem. The Naïve Bayes classifier is useful for training models on large datasets because of its simplicity. It is most suitable for categorical input features. For numerical input features, you can manually create categorical variables using WOE or other encoding techniques. You can also feed numerical variables without transformation, but the algorithm assumes that the distribution of the variable is normal (you have to be cautious about this assumption). One flip side of this classifier is the *zero frequency* problem. When you have a new category in test/score data that is not present in training data, the model will output the probability as 0. In these situations, you can use a smoothing technique. PySpark supports additive smoothing (also called Laplace smoothing) to smooth categorical data. The default value for smoothing is 1. When the value is set to 0, no smoothing will be performed. Another disadvantage of using this classifier is that it expects non-negative feature values. When your dataset is completely numeric and has negative values, it is better to standardize your data before using this classifier or to choose another classifier. The Naïve Bayes classifier is more suitable for classifying documents (spam/not spam) or multi-class problems.

PySpark Code

```
target_variable_name = "y"
nonneg_df = data.select(['age', 'day', 'duration', 'campaign', 'previous',
'y'])
#exclude target variable and select all other feature vectors
features_list = nonneg_df.columns
#features_list = char_vars #this option is used only for ChiSqselector
features_list.remove(target_variable_name)
# apply the function on our DataFrame
nonneg_df = assemble_vectors(nonneg_df, features_list, target_variable_
name)
# fit Naïve Bayes model
from pyspark.ml.classification import NaiveBayes
clf = NaiveBayes(featuresCol='features', labelCol='y')
clf_model = clf.fit(nonneg_df)
```

Regularization

Regularization is a technique used to avoid overfitting. In general, there are three types of regularization

- L1 Regularization or Lasso

- L2 Regularization or Ridge

- Elastic-net, combination of both L1 and L2

Parameters

elasticNetParam is denoted by α (alpha). regParam is denoted by λ (lambda). The regularization term is provided here:

$$\alpha\left(\lambda w_1\right)+\left(1-\alpha\right)\left(\frac{\lambda}{2}w_2^2\right), \alpha \in [0,1], \lambda \geq 0$$

The first component in the equation is the L1 or Lasso term, and the second component is the L2 or Ridge term. When `elasticNetParam` is set to 1, it performs Lasso regularization, and when it is set to 0, it performs Ridge regularization. When the value is somewhere in between 0 and 1, it performs elastic-net regularization.

Lasso or L1 Penalty

- Lasso penalty takes the absolute value of the slope and performs regularization.

- Lasso excludes variables that are not useful for the model. This can be used for feature selection.

Ridge or L2 penalty

- Ridge penalty takes the squared value of the slope and performs regularization.

- Ridge diminishes the coefficients of variables that are not useful for the model.

These techniques are useful when you are trying to evaluate a bunch of hyperparameters to select the best hyperparameters and to avoid overfitting.

Summary

- We went through supervised techniques in detail.

- We discussed various error functions and activation functions.

- We got to take a look at different optimizers available in the ML field.

- We explored a variety of supervised algorithms and saw use cases for both regression and classification.

- Finally, we touched upon regularization and how to use it to avoid overfitting.

Great job! In the next chapter, we will learn about model validation. Keep learning and stay tuned.

CHAPTER 6

Model Evaluation

"All models are wrong, but some are useful"

— George E.P. Box.

Many people try to develop models to perform a certain task (for example, predicting house prices). Often times, these models cannot represent 100 percent of reality. In our example, we cannot exactly predict a house price all the time. However, it does not mean that our model is garbage. In general, all statistical and machine learning models face this problem. Then, why build one in the first place? Even though we cannot represent reality 100 percent, we can still model useful behavior and represent reality closely enough. In our example, we can use demographic information like zip code to predict price, and this model can perform better than randomness. This lays the foundation for this entire chapter. You need to make sure that this idea is planted well, because everybody cares about the insight that you bring to the table based on your model.

That being said, you can build one model or *n* models to predict house price. Which one would you pick and what is the reason? This chapter will lay out tips and tricks for you to answer these questions. Before we dive into the model evaluation metrics, we need to cover some related concepts.

Model Complexity

Model complexity measures how well your model generalizes the information it has learned from training. In the previous chapter, we split the data into training and test. Then, we built our model on the training dataset (model fit) and applied the fitted model to the test dataset (model generalization). The whole process is shown in Figure 6-1. As an example, we used three classification models. The same concept applies when you compare models generated using different hyperparameters and several epochs of the same algorithm—like in neural networks.

© Ramcharan Kakarla, Sundar Krishnan and Sridhar Alla 2021
R. Kakarla et al., *Applied Data Science Using PySpark*, https://doi.org/10.1007/978-1-4842-6500-0_6

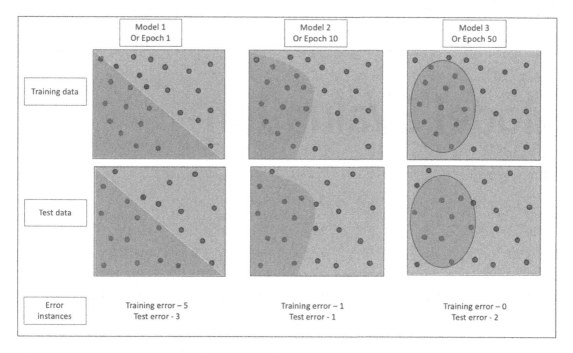

Figure 6-1. *Comparing performance of models with different settings*

Underfitting

Let's look at Model 1 in Figure 6-1. It is a linear model that does not fit our training data well. Therefore, it does not generalize well on test data either. The model fails to capture the non-linearity in the underlying data. As a result, the error level in both the training and the test datasets is high. This is called underfitting. When this type of issue happens, the model performs poorly on both the training and the test data.

Best Fitting

Now look at the Model 2 in the figure. It captures the non-linearity in the data better. As a result, the model performs equally well with both training and test data. Both our models fit, and model generalization is good. This is a best-fitting-model scenario. As a data scientist or machine learning enthusiast, you should always look for this sweet spot—the Goldilocks spot.

Overfitting

Let's look at Model 3 in the figure. This model fits the training data perfectly. However, it fails to generalize on test data well. This is called overfitting. You can also overfit a model by setting a larger number of epochs during the training process. As a result, the model memorizes (instead of generalizes) the training data well and fails to perform equally well on the test data.

When you plot the training and testing errors for each model, you typically get the graph seen in Figure 6-2. This figure is generated based on the number of error instances in the previous figure.

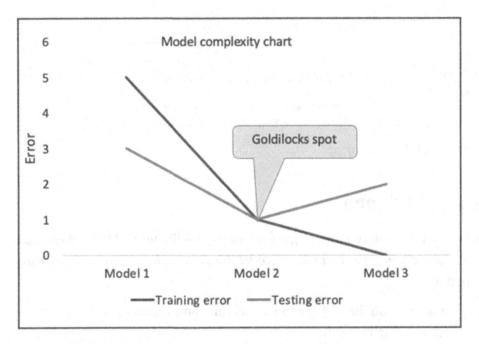

Figure 6-2. *Model complexity chart*

Often times, we would like to see errors in terms of a percentage of the whole dataset—i.e., the misclassification rate. Let's generate another figure with the misclassification rate, and this time we will use the error level at each epoch instead of using different models. You can clearly see the difference in the misclassification rates as a result of underfitting, best fit, and overfitting. By looking at this chart, you can pick the best model for your data, which is the Goldilocks spot as shown in Figure 6-3. At Epoch 10, our model performs better on both training and test data, as shown in Figure 6-3.

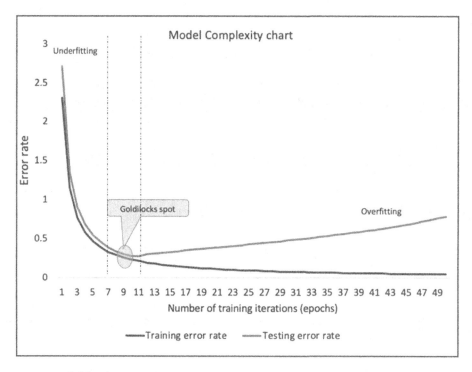

Figure 6-3. *Goldilocks spot*

Bias and Variance

When we discuss model complexity, some people use the terms *bias* and *variance* to describe it as well. *In general, an ideal model should have low bias and low variance.* So, what are they?

- *Bias* tells you the ability of an algorithm to represent the true relationship in the data.

- *Variance* tells you the consistency of an algorithm (performance) across different datasets.

In our example, Model 1 is not able to capture the non-linear relationship in the data. Therefore, it has high bias. However, it produces similar errors in both the training and the test datasets. Therefore, it has low variance. In a similar fashion, Model 2 has low bias and low variance and Model 3 has low bias and high variance. As a result, we selected Model 2 as the best model.

Model Validation

Based on what we have discussed about model complexity, we can understand the purpose of having a train/test split during modeling. Having training and testing datasets is one way to perform model validation. There are other methods available as well. But first, let's start with the train/test split and then go on to other validation techniques.

Train/Test Split

The idea is simple, as we have seen all along. You randomly split your data according to a specified size. In most of our examples, we use a 70/30 split. This means that 70 percent of the data is roughly sampled from the original dataset to form the training set, and the remaining 30 percent goes to the test set. This is shown in Figure 6-4.

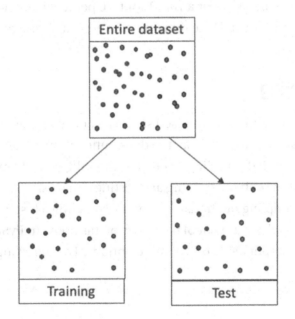

Figure 6-4. *Train/test split*

Another way to represent this method would be by using the bars seen in Figure 6-5.

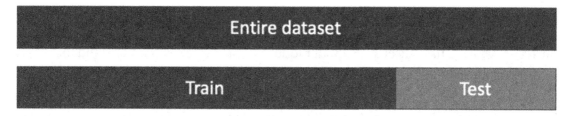

Figure 6-5. *Train/test split*

When performing the split, you can do a simple random sampling or a stratified sampling.

Simple Random Sampling

Data points are assigned at random to each train and test dataset according to the proportion specified by the user. For a 70/30 split, 70 percent of data is randomly chosen to create the training dataset, and the remaining 30 percent is assigned to the testing dataset.

Stratified Sampling

This type of sampling is done in supervised studies. Look at the distribution of the target in the original dataset. Let's assume that the distribution is 60 percent zeros and 40 percent ones. With stratified sampling, even after the split, you still maintain the 60/40 distribution of targets in both the training and testing datasets.

There are other sampling methods—like equal sampling, cluster sampling, and systematic sampling—which are out of the scope of this book. In PySpark, you can accomplish a train/test split using one of three options. The following code performs a 70/30 split.

Option 1

```
train, test = df.randomSplit([0.7, 0.3], seed=12345)
```

Option 2

The second option uses the PySpark *ML tuning* library. You need to use the vector assembler output as the input before this step.

```
from pyspark.ml.evaluation import BinaryClassificationEvaluator
from pyspark.ml.classification import LogisticRegression
from pyspark.ml.tuning import ParamGridBuilder, TrainValidationSplit

#model initialization
lr = LogisticRegression(maxIter=10, featuresCol='features',
labelCol='label')

#model parameters to try
paramGrid = ParamGridBuilder().addGrid(lr.regParam, [0.1, 0.01]).
addGrid(lr.elasticNetParam, [0.0, 0.5, 1.0]).build()

# 70% of the data will be used for training, 30% for validation.
train_valid_clf = TrainValidationSplit(estimator=lr,
estimatorParamMaps=paramGrid, evaluator=BinaryClassificationEvaluator(),
trainRatio=0.7)

# assembled_df is the output of the vector assembler
model = train_valid_clf.fit(assembled_df)
```

Option 3

To perform stratified sampling in PySpark, you need to use a workaround using the randomSplit option. The code is shown here:

```
# split data for 0s and 1s
zero_df = df.filter(df["label"]==0)
one_df = df.filter(df["label"]==1)

# split data into train and test
train_zero, test_zero = zero_df.randomSplit([0.7,0.3], seed=12345)
train_one, test_one = one_df.randomSplit([0.7,0.3], seed=12345)

# union datasets
train = train_zero.union(train_one)
test = test_zero.union(test_one)
```

Sampling Bias

We have seen the benefits of using a train/test split. However, there are some drawbacks with this approach. Imagine you are predicting some retail sales and are using November's data. Due to Thanksgiving, the store put out a huge discount, and sales increased by *x* percent. When you use this data to build a model and predict sales for the rest of the month, you might get a model that does not account for the holiday discount. This is called a sampling bias.

A sampling bias happens when your data is skewed toward one segment of the population and does not account for overall segments in the population. As a modeler, you need to understand the data before building your model. You should also know the circumstances under which your model can be used. Otherwise, the model will not produce expected results. Sampling bias could be minimized by using one of these techniques: holdout/out of time or cross-validation.

Holdout/Out of time

The holdout concept extends the idea of the train/test split, except that this time you have three datasets instead of two. See Figure 6-6.

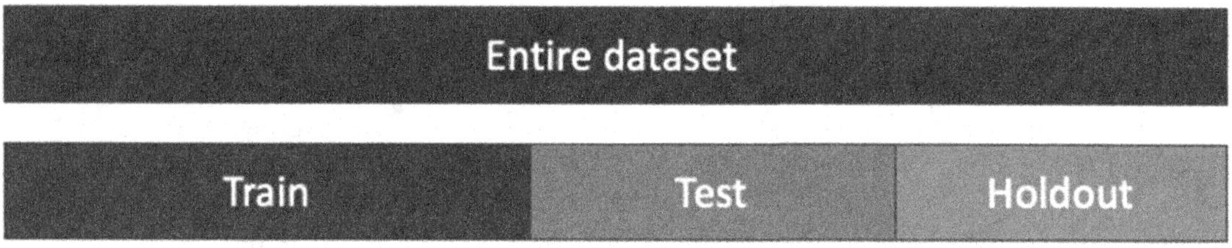

Figure 6-6. *Holdout concept*

To create a holdout dataset in PySpark, you might have to make a slight modification to the code that we saw before. This will produce a 70/20/10 split.

```
train, test, holdout = df.randomSplit([0.7, 0.2, 0.1], seed=12345)
```

In addition, you can have a training and a testing dataset before you execute Option 2, and fit the model on the training dataset instead of on the entire dataset. This way, the testing dataset can become the holdout dataset.

The holdout dataset partially solves the sampling bias issue, but the data still comes from the same timeframe. To get around this, you need to consider out-of-time data. In the retail sales prediction example, November would be training/testing/holdout, and then some other month (for example, June) would be out-of-time data. *Industry practice recommends you have at least two out-of-time data samples to validate your model.* This way, you can account for the holiday discount's significance in your model and make your predictions stable.

Population Stability Index

Another recommendation for avoiding sampling bias and to achieve a stable model would be to perform a population stability index (PSI) check on the variables across datasets (train/test/holdout/out of time). Essentially, you are measuring the distribution changes in each variable with respect to the target across your data. You pick the best variables with high feature importance and low variability (low PSI index). This way, your model produces stable predictions over time. The PSI check is a simple formula:

$$PSI = \sum \left(Actual\% - Expected\% \right) * ln\left(\frac{Actual\%}{Expected\%} \right)$$

This equation should resemble the Information Value equation we saw in the Weight of Evidence section in Chapter 4.

$$IV = \sum \left(Event\% - Non\,Event\% \right) * ln\left(\frac{Event\%}{NonEvent\%} \right)$$

The only difference between an IV and a PSI is that the IV value compares the distribution of an event and non-event of the same variable, whereas the PSI compares the distribution of the event for a variable from two different time periods (for example, training versus out of time). Therefore, the training data event distribution is the *Expected* outcome and the rest of the dataset's event distribution is considered the *Actual* outcome. For IV value, you pick the variables with high IV values. For PSI, you pick the variables with low PSI values. The cut-off for PSI is slightly different from that for IV too. See Table 6-1.

Table 6-1. *PSI Values*

PSI Value	Inference	Action
Less than 0.1	Low variability	Use this variable
0.1 - 0.25	Medium variability	Perform some transformation to make the variable stable. Cautiously, use this variable.
Greater than 0.25	High Variability	Important to address the variable stability issue before using it in model. Consider alternative variables, if stability cannot be achieved.

Let's do a simple PSI calculation for the training and holdout datasets for a single variable. See the results in Table 6-2.

Note   Run WOE code on the training dataset and apply the bins on the holdout dataset. Always the **training** dataset will be the **expected** values.

Table 6-2. *Training Versus Holdout Datasets*

varname	Value	Expected % (Train Event distribution)	Actual % (Holdout Event distribution)	Acutal % - Expected %	In (Actual%/Expected%)	Index
month	jan	0.102822581	0.097323601	-0.00549898	-0.054963466	0.00030224
month	feb	0.166931638	0.165354331	-0.001577307	-0.009493742	1.4975E-05
month	mar	0.520833333	0.517730496	-0.003102837	-0.005975263	1.854E-05
month	apr	0.187592319	0.217536071	0.029943752	0.148093588	0.00443448
month	may	0.066604128	0.068552253	0.001948126	0.028829722	5.6164E-05
month	jun	0.100320171	0.106716886	0.006396716	0.061812627	0.0003954
month	jul	0.091519731	0.089629282	-0.001890449	-0.020872517	3.9458E-05
month	aug	0.114146793	0.101141079	-0.013005714	-0.120968918	0.00157329
month	sep	0.465517241	0.462427746	-0.003089496	-0.006658815	2.0572E-05
month	oct	0.431034483	0.453703704	0.022669221	0.051256257	0.00116194
month	nov	0.108976727	0.085510689	-0.023466038	-0.242492962	0.00569035
month	dec	0.456953642	0.492063492	0.03510985	0.07402581	0.00259904
					PSI	0.016306

The PSI value for the *month* variable is 0.016. It indicates that the variable is stable (low variability) between the training and holdout datasets. You can perform the same check for all the variables in your model and make sure they are stable across both datasets. On a side note, you can do PSI checks for model outputs and monitor the performance of a model across time periods.

k-fold Cross-Validation

Until now, we have seen an approach of holdout/out of time that can handle sampling bias. What if you are in a situation where you have to work with a single dataset, and thus you don't have an out-of-time dataset? This type of situation arises especially when you deal with survey data. K-fold cross-validation could assist you in these situations. The idea is simple.

- Split the data randomly into k folds (typically the value is 5 or 10, a configurable parameter).

- Fit the model on $k-1$ folds and test the model on the left-out fold (kth fold).

- Measure the metrics (accuracy/errors) for this model.

- Repeat the process each time until all the folds are used as the test dataset.

- The final metric (accuracy/errors) is the average of all the k-fold metrics.

A pictorial representation of the k-fold (four-fold) cross-validation is shown in Figure 6-7.

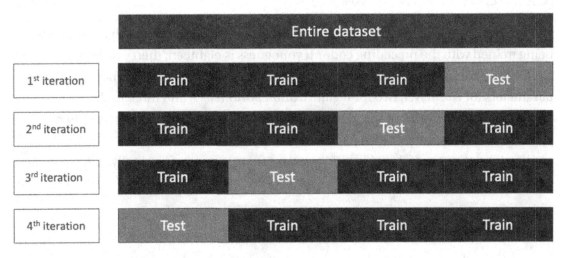

Figure 6-7. *Four-fold cross-validation*

In a k-fold cross-validation, every datapoint is used for both training and testing the model. At a given point in time, each datapoint could be either in training or testing, but not both. Therefore, the training and testing data are not static, and thus the sampling bias is minimized.

In PySpark, you can perform k-fold cross-validation using the *ML tuning* library.

```
from pyspark.ml.evaluation import BinaryClassificationEvaluator
from pyspark.ml.classification import LogisticRegression
from pyspark.ml.tuning import ParamGridBuilder, CrossValidator

#model initialization
lr = LogisticRegression(maxIter=10, featuresCol='features',
labelCol='label')

#model parameters to try
paramGrid = ParamGridBuilder().addGrid(lr.regParam, [0.1, 0.01]).
addGrid(lr.elasticNetParam, [0.0, 0.5, 1.0]).build()

# number of folds = 3
crossval_clf = CrossValidator(estimator=lr, estimatorParamMaps=paramGrid,
evaluator=BinaryClassificationEvaluator(), numFolds=3)

# assembled_df is the output of the vector assembler
model = crossval_clf.fit(assembled_df)
```

The code does a three-fold cross-validation. Can you guess the number of models being trained with the preceding code? If your guess is eighteen, then you are right. Notice the parameter grid we used to train the cross-validation model. There are three parameters for *elasticNet*, two for *regularization parameters*, and three *folds* (3 * 2 * 3 = 18 models). As you can see, cross-validation quickly becomes *expensive* when you increase the number of grid parameters or the number of folds. However, this approach becomes useful for *hyperparameter tuning* instead of manually trying out each parameter setting.

Leave-One-Out Cross-Validation

Leave-one-out cross-validation is a variation of k-fold cross-validation. The idea is to use a single observation as the test set and the rest of the data as a training set. Figure 6-8 shows the pictorial representation of this method.

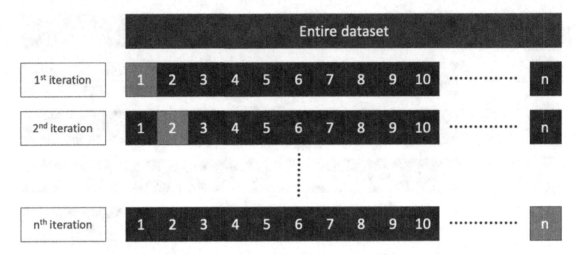

Figure 6-8. *Leave-one-out cross-validation*

Here, n is the number of observations in the data. In PySpark, you can perform leave-one-out cross-validation using the size of the dataset (n) as the number of folds in the k-fold cross-validation code shown before. This method is computationally expensive as the model needs to be trained n times for one parameter setting.

Leave-One-Group-Out Cross-Validation

This is another variant of k-fold cross-validation. It becomes useful when you want to test the model performance across different segments of people. Let's say you are doing a demographic study and want to see which age group would respond favorably to a marketing campaign. A k-fold method cannot be used here, because it does not account for the age-group segments in each fold. You want to keep your age group clean while validating the model's performance. This is where leave-one-group-out cross-validation comes into play. A pictorial representation of this method is shown in Figure 6-9 with four groups.

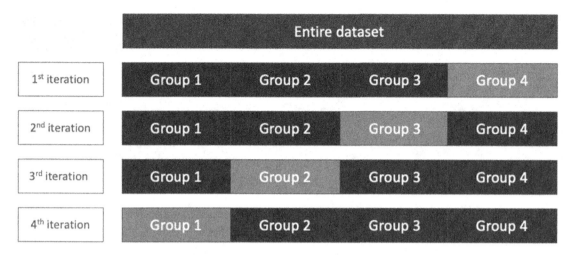

Figure 6-9. *Leave-one-group-out cross-validation*

This method is not currently available in PySpark. We have showcased a workaround code here:

```
from pyspark.sql import SparkSession
from pyspark.ml.feature import VectorAssembler
from pyspark.ml import Pipeline
from pyspark.sql.functions import countDistinct
from pyspark.ml.classification import LogisticRegression
from pyspark.ml.evaluation import BinaryClassificationEvaluator
from pyspark.sql import functions as F
import numpy as np

filename = "bank-full.csv"
spark = SparkSession.builder.getOrCreate()
df = spark.read.csv(filename, header=True, inferSchema=True, sep=';')
df = df.withColumn('label', F.when(F.col("y") == 'yes', 1).otherwise(0))
df = df.drop('y')
df = df.select(['education', 'age', 'balance', 'day', 'duration',
'campaign', 'pdays', 'previous', 'label'])
features_list = ['age', 'balance', 'day', 'duration', 'campaign', 'pdays',
'previous']
```

```
#assemble individual columns to one column - 'features'
def assemble_vectors(df, features_list, target_variable_name, group_
variable_name):
    stages = []
    #assemble vectors
    assembler = VectorAssembler(inputCols=features_list,
    outputCol='features')
    stages = [assembler]
    #select all the columns + target + newly created 'features' column
    selectedCols = [group_variable_name, target_variable_name, 'features']

    #use pipeline to process sequentially
    pipeline = Pipeline(stages=stages)
    #assembler model
    assembleModel = pipeline.fit(df)
    #apply assembler model on data
    df = assembleModel.transform(df).select(selectedCols)

    return df

# apply the function on our DataFrame
joined_df = assemble_vectors(df, features_list, 'label', 'education')
# find the groups to apply cross validation
groups = list(joined_df.select('education').toPandas()['education'].
unique())

# leave-one-group-out validation
def leave_one_group_out_validator(df, var_name, groups):

    train_metric_score = []
    test_metric_score = []

    for i in groups:
        train = df.filter(df[var_name]!=i)
        test = df.filter(df[var_name]==i)

        #model initialization
        lr = LogisticRegression(maxIter=10, featuresCol='features',
        labelCol='label')
```

```
    evaluator = BinaryClassificationEvaluator(labelCol='label',
    rawPredictionCol='rawPrediction',
                                        metricName='areaUnderROC')
    #fit model
    lrModel = lr.fit(train)
    #make predicitons
    predict_train = lrModel.transform(train)
    predict_test = lrModel.transform(test)
    train_metric_score.append(evaluator.evaluate(predict_train))
    test_metric_score.append(evaluator.evaluate(predict_test))
    print(str(i) + " Group evaluation")
    print(" Train AUC - ", train_metric_score[-1])
    print(" Test AUC - ", test_metric_score[-1])

print('Final evaluation for model')
print('Train ROC', np.mean(train_metric_score))
print('Test ROC', np.mean(test_metric_score))
```

Time-series Model Validation

When it comes to time-series validation, the train/test split is done a little differently from what we have seen so far. Let's say you are working with monthly time-series data for the past year, and you want to perform a 70/30 split. You should pick the first eight months of data for training and the next four months of data for testing. K-fold cross-validation could still be used here; however, the training data should always happen before the testing data in all the folds.

Leakage

Another common problem in machine learning is leakage, which happens when you train a model using information that is not available at the time of prediction. Leakage can occur in both the data and the target. In either case, the model will not perform as expected when deployed in production.

Target Leakage

Let's use a simple example to understand this issue. Problem Statement: *"Predict the customers who are likely to buy a product in the upcoming month."* You are using July's data, and ideally your target should come from August for this case. Instead, you create the target based on July. In this case, you are predicting the event after it happened. This does not solve the problem statement defined and is an example of target leakage.

Another variant of target leakage could happen when you use an input feature to derive the target variable. Let's say you want to predict whether a house will sell above one million dollars. You derive the target from the sales price (above 1 million dollars – 1 and below 1 million dollars – 0). In this case, target leakage happens when you include the sales price variable in the model training, since it is a proxy for the target. When target definition is complex, this problem is more likely to occur.

Data Leakage

This type of leakage is more common than target leakage. Let's assume you have a 70/30 (train/test) split. Instead of training the model on the 70 percent training data, you train the model on the entire dataset (100 percent). In this situation, the model has already seen the test data. As a result, the evaluation metrics of the model are biased.

Another variant of this issue occurs when you bootstrap the dataset before the train/test split. Bootstrapping is a concept by which you replicate the datapoints randomly. It is also known as random sampling with replacement. When you split the data after bootstrapping, it is more likely that the same datapoint is present in both datasets, thereby causing data leakage. In time-series data, when you split the data randomly instead of following the criteria (training happens before testing), it results in data leakage.

Issues with Leakage

Target or data leakage could result in significant issues, as follows:

1) Model could not be deployed because the data is not available.

2) Model performance metrics are biased.

3) The model outputs are unstable and therefore the model fails to work in production.

Proper business understanding, feature engineering, evaluation of your data/target definitions with business stakeholders, and out-of-time datasets could help you resolve leakage issues.

Model Assessment

Up until now, we have discussed the related concepts. Now, it's time to dive deep into model assessment metrics. Depending upon the type of the model, PySpark offers you a variety of model assessment metrics, as shown in Figure 6-10.

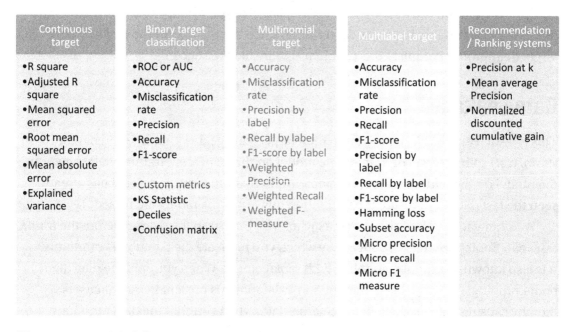

Continuous target	Binary target classification	Multinomial target	Multilabel target	Recommendation / Ranking systems
•R square	•ROC or AUC	•Accuracy	•Accuracy	•Precision at k
•Adjusted R square	•Accuracy	•Misclassification rate	•Misclassification rate	•Mean average Precision
•Mean squared error	•Misclassification rate	•Precision by label	•Precision	•Normalized discounted cumulative gain
•Root mean squared error	•Precision	•Recall by label	•Recall	
•Mean absolute error	•Recall	•F1-score by label	•F1-score	
•Explained variance	•F1-score	•Weighted Precision	•Precision by label	
	•Custom metrics	•Weighted Recall	•Recall by label	
	•KS Statistic	•Weighted F-measure	•F1-score by label	
	•Deciles		•Hamming loss	
	•Confusion matrix		•Subset accuracy	
			•Micro precision	
			•Micro recall	
			•Micro F1 measure	

Figure 6-10. *Model assessment metrics*

Continuous Target

You can evaluate continuous targets using the `RegressionEvaluator` option available in the PySpark *ML Tuning* package. Let's look at the sample data.

x1	x2	target
58	50	12
37	95	27
29	137	39
19	150	45

Figure 6-11 shows the plot of the data with the regression trendline (fit). We used Excel to produce the plot. You will get the same output using the PySpark LinearRegression model shown in the figure.

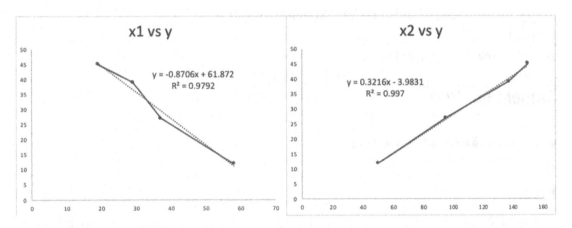

Figure 6-11. *Regresion evaluation*

We will use *x1 vs y* analysis to go through the entirety of the regression evaluation metrics.

Code to Replicate the Output Using PySpark

```
from pyspark.sql.types import IntegerType
from pyspark.sql.types import StructField, StructType
cSchema = StructType([StructField("x1", IntegerType())\
                ,StructField("x2", IntegerType())\
                ,StructField("y", IntegerType())])
df_list = [[58, 50, 12], [37, 95, 27], [29, 137, 39], [19, 150, 45]]
```

```
df = spark.createDataFrame(df_list, schema=cSchema)
# vector assembler we have used in earlier chapters
assembled_df = assemble_vectors(df, ['x1'], 'y')

#Build regression model
from pyspark.ml.regression import LinearRegression
reg = LinearRegression(featuresCol='features', labelCol='y')
reg_model = reg.fit(assembled_df) # fit model
# print coefficient and intercept
print(reg_model.coefficients[0], reg_model.intercept)
# Output: -0.8705560619872369 61.87237921604372
#prediction result
pred_result = reg_model.transform(assembled_df)
# model summary
reg_summary = reg_model.summary
```

Model equation

y-hat = -0.8706* x1 + 61.8723

Error

Error is defined as the difference between prediction and actual target. We will use the following equations to predict *y-hat* values:

```
Obs1 = -0.8706 * 58.0 + 61.8723 = 11.38
Obs2 = -0.8706 * 37.0 + 61.8723 = 29.66
Obs3 = -0.8706 * 29.0 + 61.8723 = 36.63
Obs4 = -0.8706 * 19.0 + 61.8723 = 45.33
```

We have the predictions now. Let's calculate the error for each prediction.

```
Error1 = 12 - 11.38 = 0.62
Error2 = 27 - 29.66 = -2.66
Error3 = 39 - 36.62 = 2.37
Error4 = 45 - 45.33 = -0.33
```

What happens when we add errors to quantify our model prediction results?

```
Total error= 0.62 - 2.66 + 2.37 - 0.33 = 0.0078
```

This does not make sense, because the error does not properly quantify our evaluation results. Positive errors cancel out the negative errors, making our error value very small. *This effect is more profound when you have outliers in your data, where one or two values could compensate for the rest of the errors.* Therefore, we need to use some other metric to quantify our errors. Table 6-3 summarizes each selection metric and how to use them to evaluate continuous model performance.

Table 6-3. *Selection Metrics*

Metric	Model Performance
Mean squared error (MSE)	Lower is better
Root mean square error (RMSE)	Lower is better
Mean absolute error	Lower is better
R-squared	Higher is better
Adjusted R-squared	Higher is better
Explained variance	Higher is better

Mean Squared Error (MSE)

Mean squared error takes the average of the squared errors. First, let's calculate the squared error. We will use the errors calculated earlier.

```
Sum Squared Error (SSE) = (0.62)² + (-2.66)² + (2.37)² + (-0.33)² = 13.21
```

Now, the mean squared error is the average of the above value.

```
Mean Squared Error (MSE) = 13.21/4 = 3.303
```

Let's do the same calculation using PySpark.

```
from pyspark.ml.evaluation import RegressionEvaluator
evaluator = RegressionEvaluator(labelCol='y', predictionCol='prediction',
metricName='mse')
```

```
evaluator.evaluate(pred_result)
# using model summary
print('Mean Squared error', reg_summary.meanSquaredError)
```

Root Mean Squared Error (RMSE)

Root mean squared error takes the square root of the mean squared error (MSE).

$$\text{RMSE} = \sqrt[2]{MSE} = \sqrt[2]{3.303} = 1.817$$

Let's do the same calculation using PySpark.

```
evaluator = RegressionEvaluator(labelCol='y', predictionCol='prediction',
metricName='rmse')
evaluator.evaluate(pred_result)
# using model summary
print('Root mean squared error', reg_summary.rootMeanSquaredError)
```

Mean Absolute Error (MAE)

Mean absolute error takes the average of the absolute value of the errors. Let's calculate the absolute error for our predictions.

```
Sum Absolute Error (SAE) = 0.62 + 2.66 + 2.37 + 0.33 = 5.989
```

Mean absolute error takes the average of this value.

```
Mean absolute error = 5.989/4 = 1.497
```

Let's do the same calculation in PySpark.

```
evaluator = RegressionEvaluator(labelCol='y', predictionCol='prediction',
metricName='mae')
evaluator.evaluate(pred_result)
# using model summary
print('Mean Absolute error', reg_summary.meanAbsoluteError)
```

R-squared (R^2)

R-squared is the coefficient of the determination of a regression model. It measures the goodness of fit; i.e., how well the model fits the data. There are various ways of representing the R-squared equation.

Using model fit

Let's say,

```
SST = SSM + SSE
where
SST = Sum of squares total
SSM = Sum of squares model
SSE = Sum of squares error
Then
R-squared = SSM/ SST = 1 - (SSE/SST)
```

Using model variance

```
var(model) = variance of model = MSE
var(mean) = Average of sum of squared error around target mean (MSM)
R squared = 1 - (var(model)/var(mean)) = 1 - (MSE/MSM)
```

Let's calculate it manually. This time, we will use the model variance method since we have already calculated MSE.

```
Mean of target (mean) = (12 + 27 + 39 + 45)/4 = 30.75
SS(mean) = (12 - 30.75)² + (27 - 30.75)² + (39 - 30.75)² + (45 - 30.75)²
var(mean) = (351.56 + 14.06 + 68.06 + 203.06)/4 = 636.75/4 = 159.19
var(model) = MSE = 3.303
R-squared = 1 - (3.303/159.19) = 0.9792
```

Note SS (mean) is the same as SST. Let's do the same calculation in PySpark.

```
evaluator = RegressionEvaluator(labelCol='y', predictionCol='prediction',
metricName='r2')
evaluator.evaluate(pred_result)
print('R squared', reg_summary.r2)
```

Explained Variance (Var)

Explained variance is the difference between the baseline model variance and the fitted model variance.

```
Explained variance (var) = var(mean) - var(model)
Var = 159.19 - 3.303 = 155.884
```

Let's do the same calculation in PySpark.

```
# option available for Pyspark 3.0 and above
evaluator = RegressionEvaluator(labelCol='y', predictionCol='prediction',
metricName='var')
evaluator.evaluate(pred_result)
# PySpark < 3.0
print('Explained Variance', reg_summary.explainedVariance)
```

Adjusted R-Squared (Adj. R^2)

R-squared is a biased estimate since the number of parameters influences the value. When you add a noisy parameter or completely random parameter, the R-squared value increases. Therefore, it is good to adjust the R-squared value based on the number of parameters in the model. The formula is provided here:

$$R_{adjusted}^2 = 1 - \frac{\left(1 - R^2\right)\left(N - 1\right)}{N - p - 1}$$

N – Number of observations (4 in our example)
p – Number of parameters (1 in our example, 'x1')
$R^2_{adjusted}$ = 1 - (((1-0.97924) *3)/ (4-1-1)) = 0.96886
Let's do the calculation using PySpark.
print('R squared', reg_summary.r2adj)

Binary Target

When it comes to having a binary target, PySpark offers BinaryClassification Evaluator and MulticlassClassificationEvaluator to evaluate model performance. These options are available in PySpark *ML Tuning* package. In addition, you can use the BinaryClassificationMetrics and MulticlassMetrics options as well. Unfortunately, PySpark does not support all the evaluation methods as in the Scala version. Sometimes, you need to use workarounds to get the metrics. We suggest you convert the predictions to a pandas dataset and use *scikit-learn* for generating the metrics. This would be the alternative approach. For this book, we are going to use the native PySpark approach. Table 6-4 provides the sample data (OR Logic Gate).

Table 6-4. *Sample Data*

x1	x2	target
0	0	0
0	1	1
1	0	1
1	1	1

Let's plot the data using both features x1 and x2 against the target. The plot is shown in Figure 6-12.

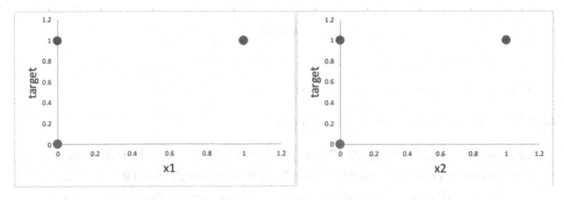

Figure 6-12. *Plot of sample data*

We will use *x1 vs target* to evaluate the classification metrics.

Code to Replicate the Output Using PySpark

```
from pyspark.sql.types import IntegerType, DoubleType
from pyspark.sql.types import StructField, StructType
cSchema = StructType([StructField("x1", IntegerType())\
                    ,StructField("x2", IntegerType())\
                    ,StructField("target", IntegerType())])
df_list = [[0, 0, 0], [0, 1, 1], [1, 0, 1], [1, 1, 1]]
df = spark.createDataFrame(df_list, schema=cSchema)
assembled_df = assemble_vectors(df, ['x1'], 'target')
from pyspark.ml.classification import LogisticRegression
clf = LogisticRegression(featuresCol='features', labelCol='target')
clf_model = clf.fit(assembled_df)
print(clf_model.coefficients[0], clf_model.intercept)
#Output: 22.453868180687905 2.0548845847848323e-06
pred_result = clf_model.transform(assembled_df)
```

Model equation

y-hat = 22.45* x1 + 0.0000021

6.4.2.3 y-hat prediction to probability to final prediction

We will use the Sigmoid function to convert our `y-hat` predictions to probabilities.

$$Sigmoid\ function, Probability = \left(\frac{1}{1+exp(-x)}\right)$$

Let us calculate predictions first using the preceding y-hat formula.

```
Obs1 = 22.45* 0 + 0.0000021 = 0.0000021
```

Similarly, we get 0.0000021, 22.45, and 22.45 for Obs2, Obs3, and Obs4, respectively. Now, let us convert our predictions to probabilities using the Sigmoid function. This results in values of 0.5, 0.5, 1.0, and 1.0, respectively. We convert the probabilities to predictions by using a cut-off value. Most of the algorithms by default use 0.5 as the cut-off. This cut-off can be modified to suit your case too. We will use 0.5 in our example as well.

```
probability >= cut-off = 1
probability < cut-off = 0
```

Based on this cut-off, we get the *final prediction* as 1, 1, 1, and 1, respectively.

Confusion Matrix

We then overlay our predictions against the actual target values to get the confusion matrix shown in Figure 6-13. The code to generate the matrix is provided for reference.

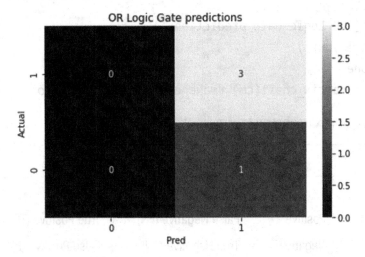

Figure 6-13. *Confusion matrix*

```
from pyspark.mllib.evaluation import BinaryClassificationMetrics,
MulticlassMetrics
# select the required columns from prediction result
predictionAndTarget = pred_result.select("prediction", "target")
predictionAndTarget = predictionAndTarget.withColumn("target",
predictionAndTarget["target"].cast(DoubleType()))
predictionAndTarget = predictionAndTarget.withColumn("prediction", predicti
onAndTarget["prediction"].cast(DoubleType()))
metrics = MulticlassMetrics(predictionAndTarget.rdd.map(tuple))
# confusion matrix
```

```
cm = metrics.confusionMatrix().toArray()
#plotting functions
import seaborn as sns
import matplotlib.pyplot as plt
def make_confusion_matrix_chart(cf_matrix):

    list_values = ['0', '1']
    plt.subplot(111)
    sns.heatmap(cf_matrix, annot=True, yticklabels=list_values,
                          xticklabels=list_values, fmt='g')
    plt.ylabel("Actual")
    plt.xlabel("Pred")
    plt.ylim([0,2])
    plt.title('OR Logic Gate predictions')
    plt.tight_layout()
    return None
make_confusion_matrix_chart(cm) #make confusion matrix plot
```

A confusion matrix in abstract terms looks like the following.

		Predicted	
		Negative	Positive
Actual	Positive	False Negative (FN)	True Positive (TP)
	Negative	True Negative (TN)	False Positive (FP)

By overlaying the abstract term matrix on our result matrix, we get the following:

```
True Positive, TP = 3
True Negative, TN = 0
False Positive, FP = 1
False Negative, FN = 0
```

Let us calculate the classification metrics one by one to evaluate this model (Table 6-5). Before that, we will go through the code to calculate them all at once.

```
from pyspark.ml.evaluation import BinaryClassificationEvaluator
acc = metrics.accuracy
misclassification_rate = 1 - acc
precision = metrics.precision(1.0)
recall = metrics.recall(1.0)
f1 = metrics.fMeasure(1.0)
evaluator = BinaryClassificationEvaluator(labelCol='target', rawPredictionC
ol='rawPrediction', metricName='areaUnderROC')
roc = evaluator.evaluate(pred_result)
print(acc, misclassification_rate, precision, recall, f1, roc)
```

Table 6-5. *Classification Metrics*

Metric	Range	Model performance	When to use it?	Examples
Accuracy	0 to 1	Higher is better.	All classes are important.	Target distribution is somewhat balanced.
Misclassification rate	0 to 1	Lower is better.	All classes are important.	Target distribution is somewhat balanced.
Precision	0 to 1	Higher is better.	Cost of false positives is high.	Scientific tests (pregnancy, COVID-19)
Recall	0 to 1	Higher is better.	Cost of false negative is high.	Imbalanced data (fraud detection)
F1-score	0 to 1	Higher is better.	Both false positives and false negatives costs are high.	Information retrieval task
ROC or AUC	0.5 to 1	Higher is better.	All classes are important.	Presentations, business stakeholders
KS Statistic	-	Higher is better.	Separation between 0 & 1 is important.	Two sample analysis

(continued)

Table 6-5. (*continued*)

Metric	Range	Model performance	When to use it?	Examples
Deciles	-	Ranking of probabilities from highest to lowest into ten segments. Ranking should not break.	Ranking of observations matters.	Marketing campaigns
Confusion matrix	-	Results (TP, TN, FP, FN)	All classes are important.	Presentations, business stakeholders
ROC curve	0.5 to 1	Performance Graph	All classes are important.	Presentations, business stakeholders
Precision recall curve	Target distribution of positive classes (for balanced classes 0.5) to 1	Performance graph	Imbalanced classes	Information retrieval task
Gini coefficient (out of scope of book)	0 to 1 Formula: (2 * AUC – 1)	Higher is better.	All classes are important.	Credit score models

Accuracy

Accuracy measures *how many times the algorithm prediction matches the actual target.* It is defined by the following formula:

$$Accuracy = \left(\frac{TP + TN}{TP + TN + FP + FN} \right)$$

The calculated accuracy value for this model is 0.75. Alternatively, you can think that one out of four predictions is wrong. Therefore, our accuracy is 0.75.

Misclassification Rate

Misclassification rate measures *how many times the algorithm's prediction did not match the actual target*. It is also known as error rate. It is defined by the following formula:

```
Misclassification rate = 1 - Accuracy
```

The misclassification rate for the model is 0.25. Alternatively, you can think that one out of four predictions is wrong. Therefore, our error rate is 0.25.

Precision

Precision measures *how many times the algorithm returns a relevant match based on its predictions*. It is also known as the positive predictive value. It is defined by the following formula:

$$Precision = \left(\frac{TP}{TP + FP} \right)$$

Our model precision is 0.75.

Recall

Recall measures *how many times the algorithm returns a relevant match based on all the relevant items*. It is also known as sensitivity or the true positive rate. It is defined by the following formula:

$$Recall = \left(\frac{TP}{TP + FN} \right)$$

Our model recall is 1.

F1-score

F1-score measures the harmonic mean between precision and recall. It is defined by the following formula:

$$F1 - score = 2 * \left(\frac{Precision * Recall}{Precision + Recall} \right)$$

Our model f1-score is 0.86.

Receiver Operating Characteristics (ROC)/Area Under the Curve (AUC)

ROC or AUC measures the area under the sensitivity and 1-specificity curve. It is shown with a light shade in Figure 6-14.

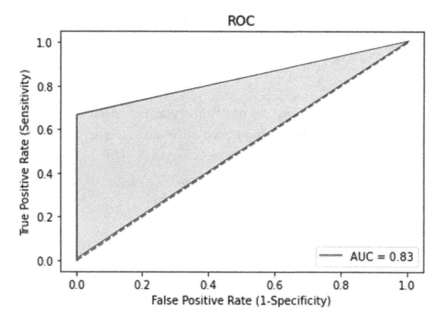

Figure 6-14. *ROC*

How is this chart generated? Well, we already know about the sensitivity (recall). Specificity is defined by the following formula:

$$Specificity = \left(\frac{TN}{TN + FP} \right) = 1 - FPR$$

where

$$False\,Positive\,Rate = \left(\frac{FP}{FP + TN} \right)$$

Remember the cut-off that we used to convert probability to prediction—we set it to 0.5 for our initial calculation. As we mentioned before, this value is customizable. Using a range of values from 0 to 1 (0, 0.1, 0.2, 0.3, 0.4, 0.5, 0.6, 0.7, 0.8, 0.9, 1) as cut-off values, we can calculate the sensitivity and the specificity using the formula.

Then, we create a graph as shown using the calculated values of sensitivity and specificity for all the ranges. For finer details in the graph, you can increase the range of values in between 0 and 1. Finally, we calculate the area under the curve formed by the graph.

The red line represents a *random model*. Since the plot is a square, the area of the triangle formed by the red line = 0.5 * base * height = 0.5 * 1 * 1 = 0.5. This means that the baseline value of ROC is 0.5. The area of the square is 1. Therefore, the maximum value of ROC is 1. To calculate the area formed by the curve (blue line) and the red line by hand, you can create squares and triangles and calculate the area of each figure separately. By adding all the areas, we get the final ROC value.

In PySpark, you can execute the following code to draw the ROC plot:

```
from pyspark.mllib.evaluation import BinaryClassificationMetrics
# calculate fpr, tpr using CurveMetrics
class CurveMetrics(BinaryClassificationMetrics):
    def __init__(self, *args):
        super(CurveMetrics, self).__init__(*args)

    def _to_list(self, rdd):
        points = []
        results_collect = rdd.collect()
        for row in results_collect:
            points += [(float(row._1()), float(row._2()))]
        return points

    def get_curve(self, method):
        rdd = getattr(self._java_model, method)().toJavaRDD()
        return self._to_list(rdd)
# use the probability colum to apply cut-offs
preds = pred_result.select('target','probability').rdd.map(lambda row:
(float(row['probability'][1]), float(row['target'])))
# Returns as a list (false positive rate, true positive rate)
points = CurveMetrics(preds).get_curve('roc')
# make plot
```

```
import matplotlib.pyplot as plt
plt.figure()
x_val = [x[0] for x in points]
y_val = [x[1] for x in points]
plt.title('ROC')
plt.xlabel('False Positive Rate (1-Specificity)')
plt.ylabel('True Positive Rate (Sensitivity)')
plt.plot(x_val, y_val, label = 'AUC = %0.2f' % roc)
plt.plot([0, 1], [0, 1], color='red', linestyle='--')
plt.legend(loc = 'lower right')
```

What Happens to ROC Curve When the Classifier Is Bad?

The curve will overlap the red line random model and thus the value of ROC will be close to 0.5.

Precision Recall Curve

The precision recall curve is the trade-off curve between precision and recall. It is useful for imbalanced classification exercises. Similar to ROC, precision and recall values are calculated for different cut-offs, and finally a plot is generated, as shown in Figure 6-15. The red line is generated based on the target distribution of 1's in the data. In our case, 50 percent of the target is 1. Therefore, the model baseline is set to 0.5.

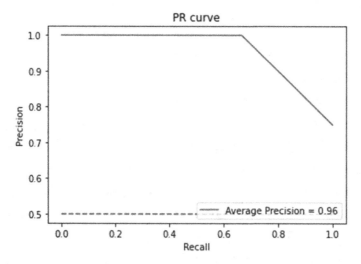

Figure 6-15. *Precision recall curve*

We can calculate the area under the blue line by averaging the precision values at different cut-offs. For this chart, it is easy to calculate the area by hand. The area of the left-out triangle on top of the figure = 0.33 * 0.25 * 0.5 = 0.0416. Therefore, the area under the PR curve = 1 - 0.0416 = 0.9584. Let's use PySpark to accomplish the same task.

```
evaluator = BinaryClassificationEvaluator(labelCol='target', rawPredictionC
ol='rawPrediction', metricName='areaUnderPR')
pr = evaluator.evaluate(pred_result)
preds = pred_result.select('target','probability').rdd.map(lambda row:
(float(row['probability'][1]), float(row['target'])))
points = CurveMetrics(preds).get_curve('pr')
plt.figure()
x_val = [x[0] for x in points]
y_val = [x[1] for x in points]
plt.title('PR curve')
plt.xlabel('Recall')
plt.ylabel('Precision')
plt.plot(x_val, y_val, label = 'Average Precision = %0.2f' % pr)
plt.plot([0, 1], [0.5, 0.5], color='red', linestyle='--')
plt.legend(loc = 'lower right')
```

What Happens to PR Curve when the Classifier Is Bad?

The area under the PR curve will resemble the area of a distribution with all target values equal to 1. In our case, the value will be close to 0.5. Let's prove that using a random value generator (Figure 6-16).

```
from sklearn.metrics import precision_recall_curve, average_precision_score
# random target and probabilities
rand_y = [random.choice([1, 0]) for i in range(0, 100)]
rand_prob = [random.uniform(0, 1) for i in range(0, 100)]
rand_precision, rand_recall, _ = precision_recall_curve(rand_y, rand_prob)
pr = average_precision_score(rand_y, rand_prob)
#plot random predictions
plt.figure()
plt.title('PR curve')
```

```
plt.xlabel('Recall')
plt.ylabel('Precision')
plt.plot(rand_recall, rand_precision, label = 'Average Precision = %0.2f'
% pr)
plt.plot([0, 1], [0.5, 0.5], color='red', linestyle='--')
plt.legend(loc = 'lower right')
```

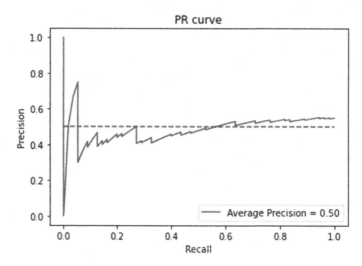

Figure 6-16. *PR curve with bad classifier*

Kolmogorov Smirnov (KS) Statistic & Deciles

The KS statistic measures the difference in the distribution of target versus non-target. It is a non-parametric evaluation. In order to perform KS, you need to first create deciles.

Deciles

To get the deciles for a model,

1) apply the model on dataset;

2) sort the scores based on descending probability (ties handled implicitly); and

3) divide the scores into ten buckets (deciles).

Well, now you have the deciles. Deciles assist you with a variety of analyses, especially in campaigns. Based on a decile, you can measure the conversion ratio in the top 10 percent, top 30 percent, and so on. In addition, you can measure lift and actual versus predicted stats.

KS Statistics

We will compute the cumulative distribution of target and non-target by decile. The difference between the two values gives you the spread of the distribution, and the maximum value of spread is the KS.

PySpark does not provide a straightforward method to perform this analysis. We have provided a workaround. For the following work, we have used the bank dataset.

```
#load dataset, cleanup and fit model
from pyspark.sql import SparkSession
from pyspark.ml.feature import VectorAssembler
from pyspark.ml import Pipeline
from pyspark.sql.functions import countDistinct
from pyspark.ml.classification import LogisticRegression
from pyspark.ml.evaluation import BinaryClassificationEvaluator
from pyspark.sql import functions as F
import numpy as np
filename = "bank-full.csv"
spark = SparkSession.builder.getOrCreate()
df = spark.read.csv(filename, header=True, inferSchema=True, sep=';')
df = df.withColumn('label', F.when(F.col("y") == 'yes', 1).otherwise(0))
df = df.drop('y')
#assemble individual columns to one column - 'features'
def assemble_vectors(df, features_list, target_variable_name):
    stages = []
    #assemble vectors
    assembler = VectorAssembler(inputCols=features_list,
    outputCol='features')
    stages = [assembler]
    #select all the columns + target + newly created 'features' column
    selectedCols = [target_variable_name, 'features']
```

```
    #use pipeline to process sequentially
    pipeline = Pipeline(stages=stages)
    #assembler model
    assembleModel = pipeline.fit(df)
    #apply assembler model on data
    df = assembleModel.transform(df).select(selectedCols)
    return df

df = df.select(['education', 'age', 'balance', 'day', 'duration',
'campaign', 'pdays', 'previous', 'label'])
features_list = ['age', 'balance', 'day', 'duration', 'campaign', 'pdays',
'previous']
assembled_df = assemble_vectors(df, features_list, 'label')
clf = LogisticRegression(featuresCol='features', labelCol='label')
train, test = assembled_df.randomSplit([0.7, 0.3], seed=12345)
clf_model = clf.fit(train)

#Deciling and KS calculation begins here
from pyspark.sql import Window

def create_deciles(df, clf, score, prediction, target, buckets):

    # get predictions from model
    pred = clf.transform(df)
    #probability of 1's, prediction and target
    pred = pred.select(F.col(score), F.col(prediction), F.col(target)).
    rdd.map(lambda row: (float(row[score][1]), float(row['prediction']),
    float(row[target])))
    predDF = pred.toDF(schema=[score, prediction, target])
    # remove ties in scores work around
    window = Window.orderBy(F.desc(score))
    predDF = predDF.withColumn("row_number", F.row_number().over(window))
    predDF.cache()
    predDF = predDF.withColumn("row_number", predDF['row_number'].
    cast("double"))
    # partition into 10 buckets
    window2 = Window.orderBy("row_number")
```

```
final_predDF = predDF.withColumn("deciles", F.ntile(buckets).
over(window2))
final_predDF = final_predDF.withColumn("deciles", final_
predDF['deciles'].cast("int"))
# create non target column
final_predDF = final_predDF.withColumn("non_target", 1 - final_
predDF[target])
final_predDF.cache()    #final predicted df

#ks calculation starts here
temp_deciles = final_predDF.groupby('deciles').agg(F.sum(target).
alias(target)).toPandas()
non_target_cnt = final_predDF.groupby('deciles').agg(F.sum('non_
target').alias('non_target')).toPandas()
temp_deciles = temp_deciles.merge(non_target_cnt, on='deciles',
how='inner')
temp_deciles = temp_deciles.sort_values(by='deciles', ascending=True)
temp_deciles['total'] = temp_deciles[target] + temp_deciles['non_
target']
temp_deciles['target_%'] = (temp_deciles[target] / temp_
deciles['total'])*100
temp_deciles['cum_target'] = temp_deciles[target].cumsum()
temp_deciles['cum_non_target'] = temp_deciles['non_target'].cumsum()
temp_deciles['target_dist'] = (temp_deciles['cum_target']/temp_
deciles[target].sum())*100
temp_deciles['non_target_dist'] = (temp_deciles['cum_non_target']/temp_
deciles['non_target'].sum())*100
temp_deciles['spread'] = temp_deciles['target_dist'] - temp_
deciles['non_target_dist']
decile_table=temp_deciles.round(2)
decile_table = decile_table[['deciles', 'total', 'label', 'non_target',
'target_%', 'cum_target', 'cum_non_target', 'target_dist', 'non_target_
dist', 'spread']]
print("KS Value - ", round(temp_deciles['spread'].max(), 2))
return final_predDF, decile_table
```

```
#create deciles on the train and test datasets
pred_train, train_deciles = create_deciles(train, clf_model, 'probability',
'prediction', 'label', 10)
pred_test, test_deciles = create_deciles(test, clf_model, 'probability',
'prediction', 'label', 10)

#pandas styling functions
from collections import OrderedDict
import pandas as pd
import sys
%matplotlib inline

def plot_pandas_style(styler):
    from IPython.core.display import HTML
    html = '\n'.join([line.lstrip() for line in styler.render().
    split('\n')])
    return HTML(html)

def highlight_max(s,color='yellow'):
    '''
    highlight the maximum in a Series yellow.
    '''
    is_max = s == s.max()
    return ['background-color: {}'.format(color) if v else '' for v in
    is_max]

def decile_labels(agg1, target, color='skyblue'):
    agg1 = agg1.round(2)
    agg1 = agg1.style.apply(highlight_max, color = 'yellow',
    subset=['spread']).set_precision(2)
    agg1.bar(subset=[target], color='{}'.format(color), vmin=0)
    agg1.bar(subset=['total'], color='{}'.format(color), vmin=0)
    agg1.bar(subset=['target_%'], color='{}'.format(color), vmin=0)
    return agg1

#train deciles and KS
plot_decile_train = decile_labels(train_deciles, 'label', color='skyblue')
plot_pandas_style(plot_decile_train)
```

	deciles	total	label	non_target		target_%	cum_target	cum_non_target	target_dist	non_target_dist	spread
0	1	3146.00	1497.00	1649.00		47.58	1497.00	1649.00	40.40	5.94	34.46
1	2	3146.00	760.00	2386.00		24.16	2257.00	4035.00	60.92	14.54	46.38
2	3	3146.00	498.00	2648.00		15.83	2755.00	6683.00	74.36	24.08	50.28
3	4	3146.00	326.00	2820.00		10.36	3081.00	9503.00	83.16	34.24	48.92
4	5	3146.00	269.00	2877.00		8.55	3350.00	12380.00	90.42	44.61	45.81
5	6	3146.00	157.00	2989.00		4.99	3507.00	15369.00	94.66	55.38	39.28
6	7	3146.00	96.00	3050.00		3.05	3603.00	18419.00	97.25	66.37	30.88
7	8	3145.00	52.00	3093.00		1.65	3655.00	21512.00	98.65	77.52	21.14
8	9	3145.00	35.00	3110.00		1.11	3690.00	24622.00	99.60	88.72	10.87
9	10	3145.00	15.00	3130.00		0.48	3705.00	27752.00	100.00	100.00	0.00

Figure 6-17. *Training dataset*

```
#test deciles and KS
plot_decile_test = decile_labels(test_deciles, 'label', color='skyblue')
plot_pandas_style(plot_decile_test)
```

	deciles	total	label	non_target		target_%	cum_target	cum_non_target	target_dist	non_target_dist	spread
0	1	1376.00	621.00	755.00		45.13	621.00	755.00	39.20	6.20	33.00
1	2	1376.00	339.00	1037.00		24.64	960.00	1792.00	60.61	14.72	45.88
2	3	1376.00	211.00	1165.00		15.33	1171.00	2957.00	73.93	24.30	49.63
3	4	1376.00	155.00	1221.00		11.26	1326.00	4178.00	83.71	34.33	49.38
4	5	1375.00	89.00	1286.00		6.47	1415.00	5464.00	89.33	44.90	44.43
5	6	1375.00	68.00	1307.00		4.95	1483.00	6771.00	93.62	55.64	37.99
6	7	1375.00	40.00	1335.00		2.91	1523.00	8106.00	96.15	66.61	29.54
7	8	1375.00	29.00	1346.00		2.11	1552.00	9452.00	97.98	77.67	20.31
8	9	1375.00	26.00	1349.00		1.89	1578.00	10801.00	99.62	88.75	10.87
9	10	1375.00	6.00	1369.00		0.44	1584.00	12170.00	100.00	100.00	0.00

Figure 6-18. *Testing dataset*

The KS value is 50.28 and 49.63 for training and testing, respectively, as shown in Figures 6-17 and 6-18. The testing KS value lies within ± 3 values from *the* train*ing* KS *value*. In addition, you can also see that the deciles do not have breaks in ranking. Therefore, this model is stable.

Actual Versus Predicted, Gains Chart, Lift Chart

Let's implement them in PySpark and then look at each metric in detail. We will use the outputs from the previous step.

```python
import matplotlib.pyplot as plt

def plots(agg1,target,type):

    plt.figure(1,figsize=(20, 5))

    plt.subplot(131)
    plt.plot(agg1['DECILE'],agg1['ACTUAL'],label='Actual')
    plt.plot(agg1['DECILE'],agg1['PRED'],label='Pred')
    plt.xticks(range(10,110,10))
    plt.legend(fontsize=15)
    plt.grid(True)
    plt.title('Actual vs Predicted', fontsize=20)
    plt.xlabel("Population %",fontsize=15)
    plt.ylabel(str(target) + " " + str(type) + " %",fontsize=15)

    plt.subplot(132)
    X = agg1['DECILE'].tolist()
    X.append(0)
    Y = agg1['DIST_TAR'].tolist()
    Y.append(0)
    plt.plot(sorted(X),sorted(Y))
    plt.plot([0, 100], [0, 100],'r--')
    plt.xticks(range(0,110,10))
    plt.yticks(range(0,110,10))
    plt.grid(True)
    plt.title('Gains Chart', fontsize=20)
    plt.xlabel("Population %",fontsize=15)
    plt.ylabel(str(target) + str(" DISTRIBUTION") + " %",fontsize=15)
    plt.annotate(round(agg1[agg1['DECILE'] == 30].DIST_TAR.
    item(),2),xy=[30,30],
            xytext=(25, agg1[agg1['DECILE'] == 30].DIST_TAR.item() + 5),
            fontsize = 13)
```

```
    plt.annotate(round(agg1[agg1['DECILE'] == 50].DIST_TAR.item(),2),
    xy=[50,50],
            xytext=(45, agg1[agg1['DECILE'] == 50].DIST_TAR.item() + 5),
            fontsize = 13)

    plt.subplot(133)
    plt.plot(agg1['DECILE'],agg1['LIFT'])
    plt.xticks(range(10,110,10))
    plt.grid(True)
    plt.title('Lift Chart', fontsize=20)
    plt.xlabel("Population %",fontsize=15)
    plt.ylabel("Lift",fontsize=15)

    plt.tight_layout()
# aggregations for actual vs predicted, gains and lift
def gains(data, decile_df, decile_by, target, score):

    agg1 = pd.DataFrame({},index=[])
    agg1 = data.groupby(decile_by).agg(F.avg(target).alias('ACTUAL')).
    toPandas()
    score_agg = data.groupby(decile_by).agg(F.avg(score).alias('PRED')).
    toPandas()
    agg1 = agg1.merge(score_agg, on=decile_by, how='inner').merge
    (decile_df, on=decile_by, how='inner')
    agg1 = agg1.sort_values(by=decile_by, ascending=True)
    agg1 = agg1[[decile_by, 'ACTUAL', 'PRED', 'target_dist']]
    agg1 = agg1.rename(columns={'target_dist':'DIST_TAR', 'deciles':
    'DECILE'})
    decile_by = 'DECILE'
    agg1[decile_by] = agg1[decile_by]*10
    agg1['LIFT'] = agg1['DIST_TAR']/agg1[decile_by]
    agg1.columns = [x.upper() for x in agg1.columns]
    plots(agg1,target,'Distribution')

# train metrics
gains(pred_train, train_deciles, 'deciles', 'label','probability')
```

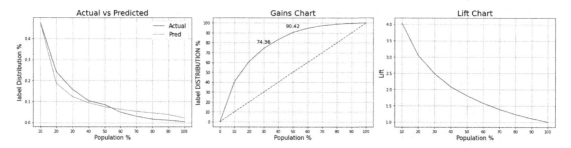

Figure 6-19a.

```
#test metrics
gains(pred_test, test_deciles, 'deciles', 'label','probability')
```

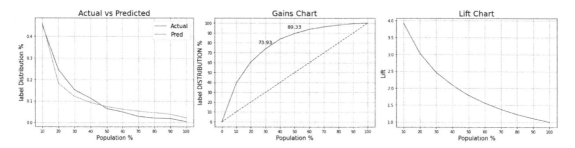

Figure 6-19b.   *Actual versus predicted, gains chart, lift chart*

Okay, we know what these metrics look like. Let's dive into each one of them.

Actual Versus Predicted

This overlays the mean distribution of targets and predictions on top of one another. When the predicted value is very close to the actual value, it signifies that the model is doing a great job, and vice-versa.

Gains Chart

This chart specifies the conversion ratio that can be achieved by targeting a certain percentage of the population. In our example here, by reaching out to the top 30 percent of customers, the conversion of targets is 74 percent. This means that almost ¾ of our targets are in the top three deciles in our model. The higher the gains chart value, the better the model. This is a very useful metric for explaining your model to the marketing team, since they focus more on conversion-based metrics.

Lift Chart

A lift chart is one more way to look at the conversion. The baseline model has a lift of 1. It means that when you target 10 percent of the population, you get 10 percent conversion. Similarly, when you target 20 percent, you get 20 percent, and so forth. However, by building a machine learning model you are now able to achieve 40 percent conversion in the top 10 percent of the population. Therefore, the lift achieved by your model in the top 10 percent of the population is four times more than the baseline model. As you go further into the deciles, the lift value reduces for your model and eventually becomes 1. Marketing teams could use this metric to understand how far they should go into deciles during campaign.

Multiclass, Multilabel Evaluation Metrics

These topics are out of scope for this book. We already used some multiclass metrics in binary model evaluation. You can apply the insights gained in this chapter to extend the metrics learned in this chapter for multiclass and multilabel classification tasks.

Summary

- We went through model complexity in detail.

- We discussed different types of model validation techniques.

- We looked at leakage in data and how to handle it for better model performance.

- Finally, we went through model evaluation metrics in detail for regression and binary models.

Great job! In the next chapter, we will learn about unsupervised techniques, topic modeling, and recommendations. Keep learning and stay tuned.

## CHAPTER 7

Unsupervised Learning and Recommendation Algorithms

This chapter will explore popular techniques used in the industry to handle unlabeled data. You will discover how you can implement them using PySpark. You will also be introduced to the basics of segmentation and recommendation algorithms and the data preparation involved therein. By end of this chapter, you will have an appreciation for unsupervised techniques and their use in day-to-day data science activities.

We will cover the following topics in this chapter:

1. K-means

2. Latent Dirichlet allocation

3. Collaborative filtering using alternating least squares

Segmentation

As the word *segmentation* suggests, it is a technique to divide your population (this could be your customers or records in your data) into logical groups that have a common need. The needs may vary highly across these logical groups, but will vary little within the groups. Let's take a step back and understand why we would need this. If you can identify a homogeneous set of customers with common needs, it helps you as a business to create a strategy to better serve those customers' needs and keep them engaged. You can create these segments to better understand your customers or for a specific motive, such as messaging or devising strategies for your application or product development.

© Ramcharan Kakarla, Sundar Krishnan and Sridhar Alla 2021
R. Kakarla et al., *Applied Data Science Using PySpark*, https://doi.org/10.1007/978-1-4842-6500-0_7

Say you are an application developer and you have data on how your application is used. It would be interesting to understand clusters of users with similar behavior. Based on how your application is used in different geographical locations, times of usage, age groups, or features used, you can drive a strategy for how you can improve the app and how you can customize the interface to different groups of users to improve engagement. Though it sounds easy, identifying the right clusters can be challenging. You wouldn't want to end up with too many clusters or very few clusters. You also would like to have a sizeable number of customers in each cluster before taking any action. You would also like these clusters to be stable. If they are changing frequently, it may pose a challenge to your intended actions.

Depending on the type of business vertical you are working with, there will be a plethora of variables available for segmentation. The question becomes, how do I choose my variables for this exercise? Before proceeding into any of the variable selection methods that were discussed earlier, it is important to identify base and descriptor variables. Base variables are those used in the segmentations for defining clusters. These could be application usage, times used, features used, etc. Descriptors are not used in the segmentations but are used to profile and define identified clusters. These descriptors could be any internal key performance indicators that you want to quantify and want to observe the differences for between clusters. For example, if we used application usage data for clustering, we would use application return rate, application churn rate, and application revenue as some of the descriptors.

Again, controlling the descriptor variables for a particular objective can yield different results. There are no hard and fast rules on what variables can be used as bases and descriptors. These are objective specific. You can also drive some of these segmentations based on survey data collected online to identify the changing needs of the customers. Data can include geographic, demographic, socioeconomic, application usage, and loyalty information. Surveys can sometimes be helpful in collecting unobservable data such as lifestyles, preferences, and benefits.

Business rules can also be used for segmenting data, but with increasing troves of data most organizations prefer using the clustering techniques available. One of the challenges with some of these techniques is that we do not know how many clusters exist in the data. The following data is made up for demonstration purposes. We are considering two factors that customers review when purchasing cars. Let's say we have two variables, safety and fuel efficiency, in this data. Both these variables were rated by fifteen respondents on a scale of 1 to 10. This data is presented in Table 7-1.

Table 7-1. *Survey Data for Car Purchase*

Respondent	Safety	Fuel Efficiency
1	10	10
2	9	9
3	9.5	9.5
4	9.75	9.75
5	8.75	8.75
6	10	1
7	9	1.5
8	9.5	2
9	8.75	1.75
10	9.25	1.25
11	1	10
12	1.5	9
13	2	9.5
14	1.75	8.75
15	1.25	9.25

If we plot the data on a scatter plot, we immediately recognize there are three clusters, as shown in Figure 7-1. For respondents in Cluster 1, safety is more important than fuel efficiency. For respondents in Cluster 2, safety and fuel efficiency are both equally important. For respondents in Cluster 3, fuel efficiency plays a major role in the purchase of vehicles. What do we do with this information? Car dealers can use this information and market the right cars to the customers. They can spend more time explaining the safety features to the respondents in Cluster 1 versus talking about fuel efficiency to the respondents in Cluster 3.

We are able to make these inferences just by plotting the variables on a two-dimensional graph. Often times we have multiple variables, and you may not be able to make these judgements visually. We will explore techniques in the next sections that we can use on multidimensional data.

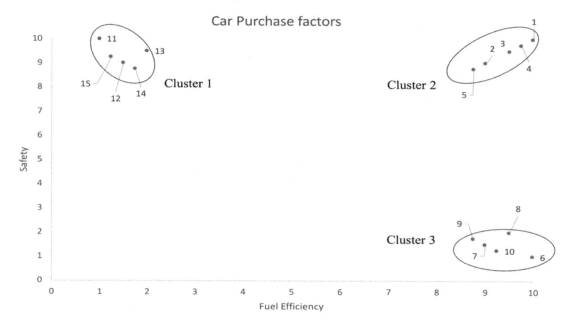

Figure 7-1. *Clusters demonstration*

There is no single clustering solution. For this example, we can classify cars based on size into compact, sedans, and SUVs, broadly. Similarly, we can classify vehicles into luxury/non-luxury and performance/non-performance segments. All these are again valid segments. Similarly, when you cluster data, there may be multiple solutions. The challenging part is to choose the one that works for your business. Descriptors can help you determine the right metrics to help you choose the solution.

Distance Measures

Based on the preceding graph, you can see that points that are close to each other have a similar characteristic. In other words, points that are separated by a small distance are more similar than points separated by larger distances. Distance between points is one of the key concepts used in segmentation for determining similarity. So, how do we calculate similarity between the points? There are multiple techniques available, one of the most popular of which is Euclidean distance. Euclidean distance is a geometric distance between two points in a cartesian coordinate system. The Euclidean distance between two points a (x_1, y_1) and b (x_2, y_2) is calculated using the following:

$$\text{Distance}_{ab} = \sqrt{(x_1 - x_2)^2 + (y_1 - y_2)^2}$$

Let's calculate this distance for Table 7-1 and see how the distances vary across the respondents. Tables 7-2, 7-3, and 7-4 represent the calculated Euclidean distances between all the respondents.

Table 7-2. *Euclidean Distance for Table 7-1 (partial output – part 1)*

	1	2	3	4	5
1	0.00	1.41	0.71	0.35	1.77
2	1.41	0.00	0.71	1.06	0.35
3	0.71	0.71	0.00	0.35	1.06
4	0.35	1.06	0.35	0.00	1.41
5	1.77	0.35	1.06	1.41	0.00
6	9.00	8.06	8.51	8.75	7.85
7	8.56	7.50	8.02	8.28	7.25
8	8.02	7.02	7.50	7.75	6.79
9	8.34	7.25	7.79	8.06	7.00
10	8.78	7.75	8.25	8.51	7.52
11	9.00	8.06	8.51	8.75	7.85
12	8.56	7.50	8.02	8.28	7.25
13	8.02	7.02	7.50	7.75	6.79
14	8.34	7.25	7.79	8.06	7.00
15	8.78	7.75	8.25	8.51	7.52

Table 7-3. *Euclidean Distance for Table 7-1 (partial output – part 2)*

	6	7	8	9	10
1	9.00	8.56	8.02	8.34	8.78
2	8.06	7.50	7.02	7.25	7.75
3	8.51	8.02	7.50	7.79	8.25
4	8.75	8.28	7.75	8.06	8.51
5	7.85	7.25	6.79	7.00	7.52
6	0.00	1.12	1.12	1.46	0.79
7	1.12	0.00	0.71	0.35	0.35
8	1.12	0.71	0.00	0.79	0.79
9	1.46	0.35	0.79	0.00	0.71
10	0.79	0.35	0.79	0.71	0.00
11	12.73	11.67	11.67	11.32	12.03
12	11.67	10.61	10.63	10.25	10.96
13	11.67	10.63	10.61	10.28	10.98
14	11.32	10.25	10.28	9.90	10.61
15	12.03	10.96	10.98	10.61	11.31

Table 7-4. *Euclidean Distance for Table 7-1 (partial output – part 3)*

	11	12	13	14	15
1	9.00	8.56	8.02	8.34	8.78
2	8.06	7.50	7.02	7.25	7.75
3	8.51	8.02	7.50	7.79	8.25
4	8.75	8.28	7.75	8.06	8.51
5	7.85	7.25	6.79	7.00	7.52
6	12.73	11.67	11.67	11.32	12.03
7	11.67	10.61	10.63	10.25	10.96
8	11.67	10.63	10.61	10.28	10.98
9	11.32	10.25	10.28	9.90	10.61
10	12.03	10.96	10.98	10.61	11.31
11	0.00	1.12	1.12	1.46	0.79
12	1.12	0.00	0.71	0.35	0.35
13	1.12	0.71	0.00	0.79	0.79
14	1.46	0.35	0.79	0.00	0.71
15	0.79	0.35	0.79	0.71	0.00

Observing the data from Tables 7-2, 7-3, and 7-4 reasserts the fact that the distance between the similar points is least. All the short distances are highlighted, and they by no means accidentally form the visual clusters shown in Figure 7-1. Well, you may have a question on how we can extend this to multiple variables. We can do it using the extension of Euclidean distance, which is extensible to any number of variables. Say we have any variable—we can calculate the distance between the points a (x_1,y_1,z_1) and b (x_2,y_2,z_2) using the following formula:

$$Distance_{ab}=\sqrt{(x1-x2)^2 +(y1-y2)^2+(z1-z2)^2}$$

There are other forms of distance measures, such as Manhattan and Chebyshev distances. For interpretation, let us consider Figure 7-2.

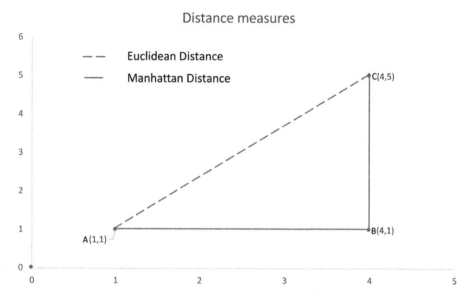

Figure 7-2. *Distance measures*

The Euclidean distance between two points *A* and *C* is 5. We can cross-verify by applying the formula. Manhattan distance between *A* and *C* is calculated as a city-block sum of the distance between (*A,B*) and (*B,C*). It is also called the taxicab metric because of its layout and is calculated for Figure 7-2 as absolute differences of cartesian coordinates.

$$\text{Manhattan distance} = |(x2 - x1)| + |(y2-y1)|$$

$$\text{Manhattan distance} = |(4{-}1)| + |(5{-}1)|$$

$$\text{Manhattan distance} = 7$$

Chebyshev distance in a vector space is defined as the distance between two vectors and is the greatest distance of their differences along any coordinate dimension. It can be calculated for Figure 7-2 using the following formula:

$$\text{Chebyshev distance} = \max\,[(x2 - x1), (y2-y1)]$$

$$\text{Chebyshev distance} = \max\,[(4 - 1), (5{-}1)]$$

$$\text{Chebyshev distance} = 4$$

This strategy works great for numerical variables, but how can we apply this technique to categorical variables? Say we have a variable called "previous purchase within 5 years" with values yes/no. How can we use this variable? Remember we introduced one hot encoding in an earlier chapter. We will be able to convert this categorical variable yes/no to a numeric variable using one hot encoding. We will end up with two variables: one for "yes" and another for "no." In practice, we may just need a single variable, as the second variable might be perfectly collinear with the first variable.

If we observe the data carefully, we realize the earlier variables were on a scale of 1 to 10, and now the new variable "previous purchase within 5 years" has a scale of 0 to 1. Using a combination of these variables will cause a bias, as the new variable will have a negligible effect when we calculate the distances. How do we approach this scenario? Yes, you can use normalization. Normalization is a scaling technique t can rescale the existing data in a range of [0,1]. It is calculated using the following formula:

$$X_{normalized} = (X-X_{min}) / (X_{max} -X_{min})$$

This is a widely used technique in segmentation. But be cautious, as you might lose information on the outliers.

Another scaling technique that is widely used is standardization. Do not confuse standardization with normalization. In standardization, we rescale data to have a mean of 0 and a standard deviation of 1. It is calculated using the following formula:

$$X_{standardized} = X- X_{mean} / X_{standard\_deviation}$$

There might be situations where you only have categorical data. Are there better techniques than one hot encoding and using Euclidean distance? The answer is yes: Jaccard's coefficient and Hamming distance. We will focus here on Jaccard's coefficient. It involves calculating the dissimilarity matrix from the data. For example, if we are dealing with binary attributes we can create a contingency table where a is the number of features that equal 1 for both items x and y; b is the number of features that equal 1 for item x but equal 0 for object y; c is the number of features that equal 0 for item x but equal 1 for item y; and d is the number of features that equal 0 for both items x and y. The total number of features is z, where $z = a + b + c + d$ (Table 7-5).

Table 7-5. *Contingency Table*

x	y			
		1	0	sum
	1	a	b	a+b
	0	c	d	c+d
	sum	a+c	b+d	z =(a+b+c+d)

There are a couple of terms you need to understand for calculating the dissimilarities.

> Symmetric Binary Variable – Both levels of the variable have equal weight. For example, heads and tails of a coin

> Asymmetric Binary Variable – The two levels do not have the same importance. For example, with results of a medical test, positive may have higher weightage than the negative occurrence

Depending on the type of binary variable, your calculations for the dissimilarity will vary. Symmetric Binary dissimilarity can be calculated using

$$d(x,y) = (b+c) / (a+b+c+d)$$

Asymmetric Binary dissimilarity can be calculated using

$$d(x,y) = (b+c) / (a+b+c)$$

Observe in the preceding equation that d is discarded in the denominator; this is due to the fact that negative occurrence has little to no importance. Using this dissimilarity metric, we can calculate the similarity distance between x and y using the following formula:

$$Similarity(x,y) = 1 - d(x,y)$$

$$Similarity(x,y) = 1 - (\,(b+c) / (a+b+c)\,)$$

$$Similarity(x,y) = c/(a+b+c)$$

This similarity measure is called Jaccard's distance. This can be used to figure out the similarity between binary variables. But what happens to nominal variables? One workaround for nominal variables is to recode them to multiple binary variables. The other technique is to use a simple matching coefficient. It is calculated as follows:

Simple matching coefficient = number of matching attributes / number of attributes

In the case of a binary variable, this coefficient is similar to symmetric binary dissimilarity. The preceding coefficient gives the dissimilarity index. Similarity distance is calculated by subtracting this coefficient from 1:

Simple matching distance = 1 - Simple matching coefficient

Let's demonstrate the Jaccard's distance using a continuation of the example we used for Euclidean distances. Say a car dealership was successful in making a deal with three respondents who owned three cars of different makes and models. Now the car dealership wants to categorize the cars of these respondents based on performance. They perform different tests on these cars to determine the similarity of current performance. Failed tests are represented by Y and passed tests are represented by N. We have converted them to 1 and 0 respectively for our convenience here (Table 7-6).

Table 7-6. *Jaccard's Demonstration*

	Climatic	Mechanical	Pressure Impulse	Emissions Bench	Cat Aging	SHED
Car 1	1	0	1	0	0	0
Car 2	1	0	1	0	1	0
Car 3	1	1	0	0	0	0

We can calculate the dissimilarity between the cars using the following formula:

$$d(car1, car2) = 1+0 / (2+1+0)$$

$$d(car1, car2) = 0.33$$

Okay, how did we arrive at these numbers? Well, using the contingency table, we counted the number of b's and c's as represented in Table 7-6 . For Car 1 and Car 2, Cat Aging accounts to b. There is no c value here. Also note this is asymmetric binary similarity, so we would not account for the value of d in the denominator. From climatic and pressure impulse tests we know the count of a is 2. To calculate the similarity we need to subtract this index from 1:

$$Similarity(car1, car2) = 1 - 0.33$$

$$Similarity(car1, car2) = 0.67$$

Similarly, you can calculate the differences between other observations and create the similarity matrix.

Types of Clustering

Broadly, we can divide clustering into hierarchical and non-hierarchical clustering. We will focus in this section on understanding the differences between these types of clustering and on the algorithms available for each type of clustering. As the name suggests, hierarchical clustering tries to build a hierarchy of clusters in its routine to finalize the solution. Hierarchical clustering is further divided into agglomerative and divisive (Figure 7-3).

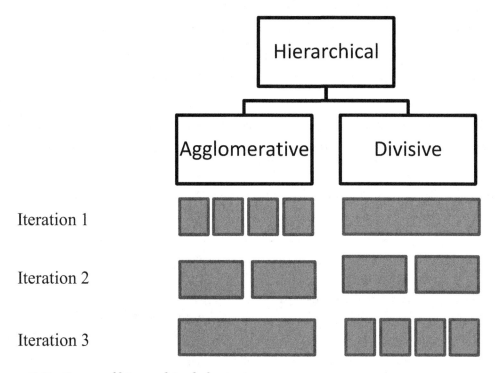

Figure 7-3. *Types of hierarchical clustering*

Agglomerative Clustering: In this type of clustering, each observation starts in its own cluster, working its way up to a single cluster. This approach is also referred to as bottom up.

Divisive Clustering: In this type of clustering, all observations are assumed to fall in a single cluster. They are then divided in each iteration recursively. This approach is also referred to as top down.

Both approaches are pictorially depicted in Figure 7-3. We will make use of selected data from Table 7-1 for demonstrating the agglomerative example. We also calculated Euclidean distance for all these points (Table 7-8). All pair-wise distances are represented in Table 7-7.

Table 7-7. *Agglomerative Example*

Respondent	Safety	Fuel Efficiency
1	10	10
2	9	9
6	10	1
7	9	1.5
11	1	10
12	1.5	9

Table 7-8. *Euclidean Distances of Selected Respondents*

	1	2	6	7	11	12
1	0.00	1.41	9.00	8.56	9.00	8.56
2	1.41	0.00	8.06	7.50	8.06	7.50
6	9.00	8.06	0.00	1.12	12.73	11.67
7	8.56	7.50	1.12	0.00	11.67	10.67
11	9.00	8.06	12.73	11.67	0.00	1.12
12	8.56	7.50	11.67	10.61	1.12	0.00

Note that this is a symmetric matrix; you can use either half to determine your clusters. When we start agglomerative clustering, all the datapoints are in their individual clusters. Based on the preceding data, the smallest distance observed between two respondents here is 1.12 for (6,7) and (11,12). There is a tie, so how will the algorithm decide which pairs should be fused? Well, there are built-in rules to break this tie, and let's say it arbitrarily picks (6,7). In the first step, we would cluster them together and then calculate the overall heterogeneity measure, which here is equal to 1.12. Note that the heterogeneity measure as we begin the clustering is 0 because all the datapoints are in their own individual clusters. Once we fuse 6 and 7 together, the next pair that gets fused is (11,12). The heterogeneity measure remains the same because the pair-wise distance between the points is still 1.12. In the next step, (1,2) will be fused together since they have the least pair-wise distance of 1.41. Now the average heterogeneity will go up to 1.21. This is calculated by taking (1.12+1.12+1.41)/3. These steps are repeated until all the observations fall into a single cluster.

You may wonder, how would I choose the right number of clusters? One way to identify this is to determine the break/fluctuation in the heterogeneity measure. Figure 7-4 gives you a visual depiction of how the algorithm works. Table 7-9 shows the steps.

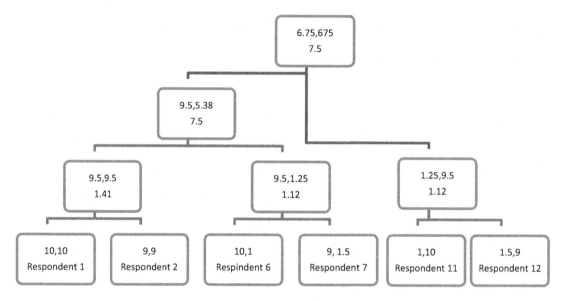

Figure 7-4. *Dendrogram: Hierarchical clustering based on single linkage. Centroids and Euclidean distances represented*

Table 7-9. *Clustering Steps*

Step	Clusters	Heterogeneity
0	1 2 6 7 11 12	0
1	1 2 (**6 7**) 11 12	1.12
2	1 2 (**6 7**) (**11 12**)	1.12
3	(**1 2**) (**6 7**) (**11 12**)	1.21
4	(**1 2 6 7**) (11 12)	3.84
5	(**1 2 6 7 11 12**)	5.04

Observing the preceding heterogeneity measure, you can observe there is break in the measure from Step 3 to Step 4 and from Step 4 to Step 5. You can arrive at two possible solutions here, either a three-cluster or a four-cluster solution.

The method we used here is called single linkage. It looks at the nearest neighbor in determining the cluster measures and heterogeneity. Other methods available for looking at similarities include the following:

Complete Linkage: This uses the maximum distance between clusters to determine the cluster distances, instead of using the minimum. See Figure 7-5 for how the cluster attributes and distances vary .

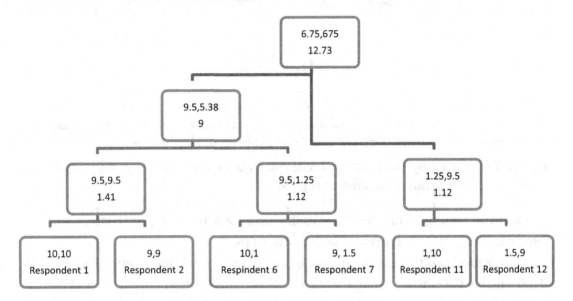

Figure 7-5. *Dendrogram: Hierarchical clustering based on complete linkage. Centroids and Euclidean distances represented*

Average Linkage: This uses the pair-wise average distance between clusters to determine the cluster distances. See Figure 7-6 for how the cluster distances vary.

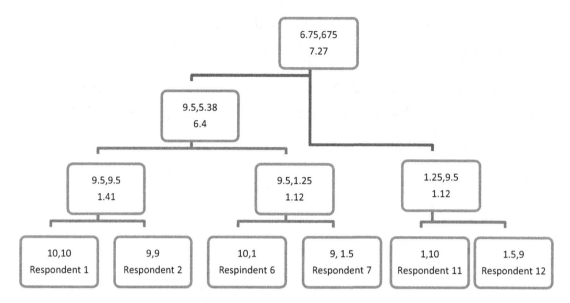

Figure 7-6. *Dendrogram: Hierarchical clustering based on average linkage. Centroids and Euclidean distances represented*

Centroid Linkage: This method uses the distance between cluster centers to determine the cluster distances in the iterative steps.

Ward's method: This method is based upon variance. It is based upon the principle of minimizing the cluster variance between observations within the cluster and maximizing the cluster variances across the clusters.

Although hierarchical clustering offers the advantage of your not having to pick the initial number of clusters, it is a complex algorithm to implement, especially for a big data environment. It has a time and space complexity order of $O(n^2\log(n))$, where n is the number of observations. This can slow operations. On the brighter side, it can provide an understanding of how the observations are grouped.

Bisecting k-Means

Bisecting k-means is a hybrid approach between hierarchical divisive clustering and non-hierarchical clustering. We will introduce k-means in the next section, which is one of the popular non-hierarchical clustering techniques. Bisecting k-means is similar to k-means in predefining the number of clusters. You may wonder how we can set the number of clusters without any prior knowledge of the data. There are the techniques we saw earlier, like the heterogeneity measures, and other distance metrics we can make use of to determine the clusters if we run the algorithm for a few iterations. Let's see how this bisecting k-means algorithm works in theory. Since this is a divisive algorithm, it starts at the top with a single cluster and then divides data iteratively. First, we need to set the required number of clusters.

As we start, all the data exists in a single cluster (Figure 7-7). We split the data into two clusters (Figure 7-8).

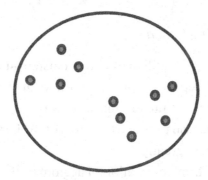

Figure 7-7.   *Step 1: Bisecting k-means*

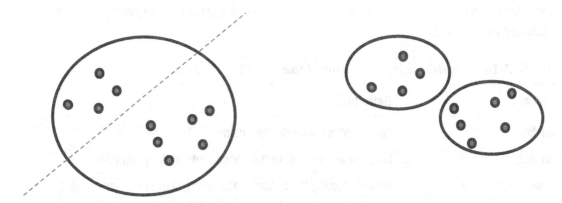

Figure 7-8.   *Step 2: Bisecting k-means*

These first two clusters are divided using k-means. As the clusters are split, the centroid of each cluster is calculated. Then, the sum of the square of the distance measure between all the points and the centroid is calculated. This is called the sum of squared distances/errors (SSE). The higher the SSE, the greater the heterogeneity of the cluster. In the next step, we choose the cluster with the highest SSE among the clusters and divide it into two clusters (Figure 7-9).

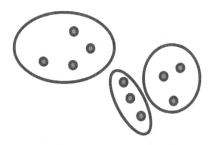

Figure 7-9. *Step 3: Bisecting k-means*

In the next step, we look at the SSE of the three clusters and choose the one with the highest SSE to bisect. We repeat the process until we arrive at the desired number of clusters. The low computation time for this process is a great advantage over other processes. Bisecting k-means also produces clusters of similar sizes, unlike k-means, where the cluster sizes can vary greatly.

PySpark doesn't have an implementation of agglomerative hierarchical clustering as of version 3.0. There is a divisive hierarchical clustering method (top-down) approach using the `BisectingKMeans` function. For demonstration purposes, we are using credit card data that is sourced from Kaggle (`https://www.kaggle.com/arjunbhasin2013/ccdata`)(Table 7-10).

Table 7-10. *Metadata Information of the Credit Card Dataset*

Column	Definition
custid	Identification of credit card holder (Categorical)
balance	Balance amount left in their account to make purchases
balancefrequency	How frequently the balance is updated, score between 0 and 1 (1 = frequently updated, 0 = not frequently updated)
purchases	Amount of purchases made from account
oneoffpurchases	Maximum purchase amount done in one go
installmentspurchases	Amount of purchase done in installments
cashadvance	Cash in advance given by the user
purchasesfrequency	How frequently purchases are being made, score between 0 and 1 (1 = frequently purchased, 0 = not frequently purchased)

(*continued*)

Table 7-10. (*continued*)

Column	Definition
oneoffpurchasesfrequency	How frequently purchases are happening in one go (1 = frequently purchased, 0 = not frequently purchased)
purchasesinstallmentsfrequency	How frequently purchases in installments are being done (1 = frequently done, 0 = not frequently done)
cashadvancefrequency	How frequently cash in advance is being paid
cashadvancetrx	Number of transactions made with "cash in advance"
purchasestrx	Number of purchase transactions made
creditlimit	Limit of credit card for user
payments	Amount of payment done by user
minimum_payments	Minimum amount of payments made by user
prcfullpayment	Percentage of full payment paid by user
tenure	Tenure of credit card service for user

Now, this dataset can't be used in its raw form. We know from the descriptions that these data elements have different scales. We will have to normalize them and form a feature vector before we apply the clustering technique.

```
# Read data
file_location = "cluster_data.csv"
file_type = "csv"
infer_schema = "false"
first_row_is_header = "true"

df = spark.read.format(file_type)\
.option("inferSchema", infer_schema)\
.option("header", first_row_is_header)\
.load(file_location)

# Print metadata
df.printSchema()

# Casting multiple variables
```

```
from pyspark.sql.types import *

#Identifying and assigning lists of variables
float_vars=list(set(df.columns) - set(['CUST_ID']))

for column in float_vars:
    df=df.withColumn(column,df[column].cast(FloatType()))

# Imputing data


from pyspark.ml.feature import Imputer


# Shortlisting variables where mean imputation is required
input_cols=list(set(df.columns) - set(['CUST_ID']))

# Defining the imputer function
imputer = Imputer(
    inputCols=input_cols,
    outputCols=["{}_imputed".format(c) for c in input_cols])

# Applying the transformation
df_imputed=imputer.fit(df).transform(df)

# Dropping the original columns as we created the _imputed columns
df_imputed=df_imputed.drop(*input_cols)

# Renaming the input columns to original columns for consistency
new_column_name_list= list(map(lambda x: x.replace("_imputed", ""),
df.columns))
df_imputed = df_imputed.toDF(*new_column_name_list)

# Data Preparation


from pyspark.ml.feature import VectorAssembler
from pyspark.ml.feature import Normalizer
from pyspark.ml import Pipeline

# Listing the variables that are not required in the segmentation analysis
ignore = ['CUST_ID']
```

```python
# creating vector of all features
assembler = VectorAssembler(inputCols=[x for x in df.columns if x not in
ignore],
                                outputCol='features')
# creating the normalization for all features for scaling betwen 0 to 1
normalizer = Normalizer(inputCol="features", outputCol="normFeatures",
p=1.0)
# Defining the pipeline
pipeline = Pipeline(stages=[assembler, normalizer])
# Fitting the pipeline
transformations=pipeline.fit(df_imputed)
# Applying the transformation
df_updated = transformations.transform(df_imputed)

# Building the model


from pyspark.ml.clustering import BisectingKMeans
from pyspark.ml.evaluation import ClusteringEvaluator
# Trains a bisecting k-means model.
bkm = BisectingKMeans().setK(2).setSeed(1)
model = bkm.fit(df_updated.select('normFeatures').withColumnRenamed('normFe
atures','features'))

# Make predictions
predictions = model.transform(df_updated.select('normFeatures').withColumnR
enamed('normFeatures','features'))

# Evaluate clustering by computing Silhouette score
evaluator = ClusteringEvaluator()

silhouette = evaluator.evaluate(predictions)
print("Silhouette with squared euclidean distance = " + str(silhouette))

# Shows the result.
print("Cluster Centers: ")
centers = model.clusterCenters()
for center in centers:
    print(center)
```

We use the Silhouette score to determine the number of clusters. This is one of the best available metrics to determine the right number of clusters. This metric is calculated for each sample in the dataset. It is calculated using the following formula:

$$\text{Silhouette Coefficient} = (x-y)/\max(x,y)$$

The x is the inter-cluster distance and y is the intra-cluster distance. The coefficient can vary between -1 and 1. Values close to 1 define a clear separation and are more desirable. Let's now calculate the Silhouette coefficient for different numbers of clusters and plot it to determine the right cluster number (Figure 7-10).

```
# Iterations

import pandas as pd
import seaborn as sns
import matplotlib.pyplot as plt

sil_coeff=[]
num_clusters=[]
for iter in range(2,8):
    bkm = BisectingKMeans().setK(iter).setSeed(1)
    model = bkm.fit(df_updated.select('normFeatures').withColumnRenamed('no
    rmFeatures','features'))
    # Make predictions
    predictions = model.transform(df_updated.select('normFeatures').withCol
    umnRenamed('normFeatures','features'))
    # Evaluate clustering by computing Silhouette score
    evaluator = ClusteringEvaluator()
    silhouette = evaluator.evaluate(predictions)
    sil_coeff.append(silhouette)
    num_clusters.append(iter)
    print("Silhouette with squared euclidean distance for "+str(iter) +"
    cluster solution = " + str(silhouette))

df_viz=pd.DataFrame(zip(num_clusters,sil_coeff), columns=['num_
clusters','silhouette_score'])
sns.lineplot(x = "num_clusters", y = "silhouette_score", data=df_viz)
plt.title('Bisecting k-means : Silhouette scores')
plt.xticks(range(2, 8))
plt.show()
```

```
Silhouette with squared euclidean distance for 2 cluster solution = 0.49855848955650006
Silhouette with squared euclidean distance for 3 cluster solution = 0.47032306895494647
Silhouette with squared euclidean distance for 4 cluster solution = 0.36841230437497996
Silhouette with squared euclidean distance for 5 cluster solution = 0.35979911858435865
Silhouette with squared euclidean distance for 6 cluster solution = 0.3787940202378902
Silhouette with squared euclidean distance for 7 cluster solution = 0.3639517765678912
```

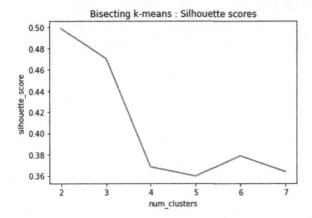

Figure 7-10. *Silhouette score changing over multiple iterations using the bisecting k-means with an increasing number of clusters*

K-means

By far, k-means is one of the most popular clustering techniques. The reason is because it scales with data. Most other techniques are computationally expensive after certain datapoints. One of the advantages and sometimes a disadvantage with k-means is that you can dictate the number of clusters that you want. For demonstration, we will use the same car survey data from Table 7-1. Let us pick the number of clusters to be three. Now the algorithm assigns three random initialization points in the space, as shown in Figure 7-11. These points represent initial cluster centers.

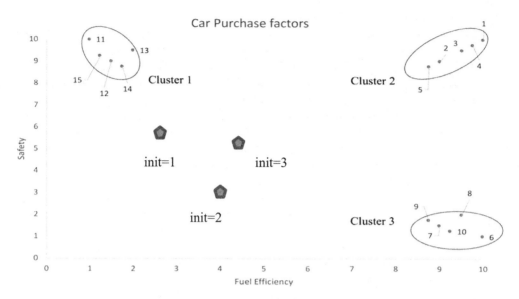

Figure 7-11. *K-means demonstration*

In the second step, the distances of all points from the three initialization-point cluster centers are calculated. Based on the minimum distance to the cluster centers, all points are assigned to one of the clusters. Centroids of new cluster points are calculated to find the new cluster center. Again, the distances of all points from the new cluster centers are calculated. Based on the minimum distance to cluster centers, points are reassigned. This process repeats until there is no change in the points in a cluster.

You may wonder, how do we identify the right number of clusters in k-means? Well, we know the answer from hierarchical clustering. We use the cluster variances to determine the right number of clusters. We run the clustering from 0 to n, plotting the change in the cluster variances. This plot is called an elbow plot. Whenever you observe an elbow or a sharp change in the plot, you can use that point as the optimal number of clusters. The goal is identifying the smallest value of k that has lowest sum of squares (SSE).

We will make use of the same credit card dataset to implement k-means in PySpark.

EXERCISE 7-1: SEGMENTATION

Question 1: Observe the differences between hierarchical and non-hierarchical clustering by setting the same number of clusters.

- Are the cluster profiles and sizes similar?

```python
# Import Sparksession
from pyspark.sql import SparkSession
spark=SparkSession.builder.appName("Clustering").getOrCreate()

# Print PySpark and Python versions
import sys
print('Python version: '+sys.version)
print('Spark version: '+spark.version)

# Read data
file_location = "cluster_data.csv"
file_type = "csv"
infer_schema = "false"
first_row_is_header = "true"

df = spark.read.format(file_type)\
.option("inferSchema", infer_schema)\
.option("header", first_row_is_header)\
.load(file_location)

# Print metadata
df.printSchema()

#  Count data
df.count()
print('The total number of records in the credit card dataset are '+str(df.
count()))

# Casting multiple variables
from pyspark.sql.types import *

# Identify and assign lists of variables
float_vars=list(set(df.columns) - set(['CUST_ID']))

for column in float_vars:
    df=df.withColumn(column,df[column].cast(FloatType()))

from pyspark.ml.feature import Imputer

# Shortlist variables where mean imputation is required
```

275

```python
input_cols=list(set(df.columns) - set(['CUST_ID']))

# Define the imputer function
imputer = Imputer(
    inputCols=input_cols,
    outputCols=["{}_imputed".format(c) for c in input_cols])

# Apply the transformation
df_imputed=imputer.fit(df).transform(df)

# Drop the original columns as we created the _imputed columns
df_imputed=df_imputed.drop(*input_cols)

# Rename the input columns to original columns for consistency
new_column_name_list= list(map(lambda x: x.replace("_imputed", ""),
df.columns))
df_imputed = df_imputed.toDF(*new_column_name_list)

from pyspark.ml.feature import VectorAssembler
from pyspark.ml.feature import Normalizer
from pyspark.ml import Pipeline

# List the variables that are not required in the segmentation analysis
ignore = ['CUST_ID']

# create vector of all features
assembler = VectorAssembler(inputCols=[x for x in df.columns if x not in
ignore],
                            outputCol='features')

# create the normalization for all features for scaling betwen 0 to 1
normalizer = Normalizer(inputCol="features", outputCol="normFeatures",
p=1.0)
# Define the pipeline
pipeline = Pipeline(stages=[assembler, normalizer])
# Fit the pipeline
transformations=pipeline.fit(df_imputed)
# Apply the transformation
df_updated = transformations.transform(df_imputed)
```

```python
from pyspark.ml.clustering import KMeans
from pyspark.ml.evaluation import ClusteringEvaluator

# Train a k-means model.
kmeans = KMeans().setK(2).setSeed(1003)
model = kmeans.fit(df_updated.select('normFeatures').withColumnRenamed('nor
mFeatures','features'))

# Make predictions
predictions = model.transform(df_updated.select('normFeatures').withColumnR
enamed('normFeatures','features'))

# Evaluate clustering by computing Silhouette score
evaluator = ClusteringEvaluator()

silhouette = evaluator.evaluate(predictions)
print("Silhouette with squared euclidean distance = " + str(silhouette))

# Show the result.
centers = model.clusterCenters()
print("Cluster Centers: ")
for center in centers:
    print(center)

import pandas as pd
import seaborn as sns
import matplotlib.pyplot as plt

sil_coeff=[]
num_clusters=[]
for iter in range(2,8):
    kmeans = KMeans().setK(iter).setSeed(1003)
    model = kmeans.fit(df_updated.select('normFeatures').withColumnRenamed(
    'normFeatures','features'))
    # Make predictions
    predictions = model.transform(df_updated.select('normFeatures').withCol
    umnRenamed('normFeatures','features'))
    # Evaluate clustering by computing Silhouette score
    evaluator = ClusteringEvaluator()
```

```
    silhouette = evaluator.evaluate(predictions)
    sil_coeff.append(silhouette)
    num_clusters.append(iter)
    print("Silhouette with squared euclidean distance for "+str(iter) +"
    cluster solution = " + str(silhouette))

df_viz=pd.DataFrame(zip(num_clusters,sil_coeff), columns=['num_
clusters','silhouette_score'])
sns.lineplot(x = "num_clusters", y = "silhouette_score", data=df_viz)
plt.title('k-means : Silhouette scores')
plt.xticks(range(2, 8))
plt.show()
```

Latent Dirichlet Allocation (LDA)

Latent Dirichlet allocation can help us identify—using statistical models—abstract topics that occur in a collection of documents. LDA is fairly recent advancement and was first proposed in 2003. There are other applications of LDA, such as sentiment analysis and object localization for images, to name a few. To understand LDA, we will start with a simple diagram illustration (Figure 7-12).

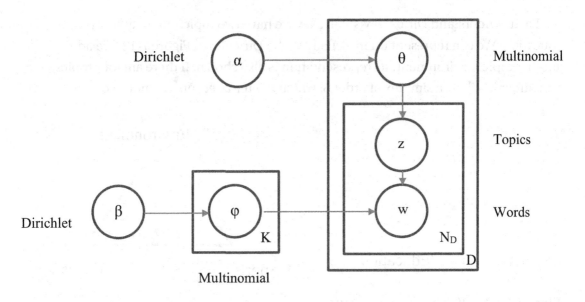

$$P(\boldsymbol{W}, \boldsymbol{Z}, \boldsymbol{\theta}, \boldsymbol{\varphi}; \alpha, \beta) = \prod_{i=1}^{K} P(\varphi_i; \beta) \prod_{j=1}^{M} P(\theta_j; \alpha) \prod_{t=1}^{N} P(Z_{j,t} \mid \theta_j) P(W_{j,t} \mid \varphi_{Z_{j,t}})$$

Figure 7-12. *Latent Dirichlet allocation demonstration*

At first glance, the diagram and the formula can be overwhelming. Let's break this down and understand the components. First, what is a Dirichlet? Dirichlet distribution is a family of continuous multivariate probability distributions parameterized by a vector of positive reals. It is a multivariate generalization of the beta. α and β in Figure 7-12 are the Dirichlet distributions. θ and φ are multinomial distributions. Using these distributions, we create words and topics, which in turn can create documents.

You can think of LDA as an engine that spits out documents. In the left-hand side of the equation we have a probability of a document's appearing. As we tune the factors on the right we can expect the generation of different flavors of documents with topics and words. The first two components of the equation are Dirichlet distributions. Words (β) can be a simplex of an order n depending upon the number of keywords. Topics (α) can be a simplex of an order less than or equal to n.

Let us understand Dirichlet visually. Say we have two topics, namely sports and education. We can represent them visually as the Dirichlet in Figure 7-13. We add another topic, environment; its representation is also shown. If there are four topics, remember it will be a simplex of order 3, which is a tetrahedron and not a square.

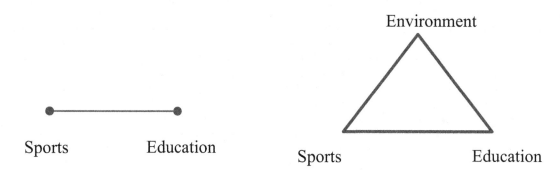

Figure 7-13. *Dirichlet representation*

If there are documents that contain sports they will be represented towards the sports side, and documents for education will be represented toward education. This is how the distributions will play out with different values of α (Figure 7-14).

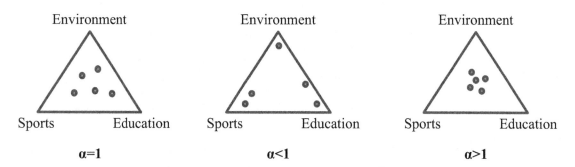

Figure 7-14. *Variations of Dirichlet distribution*

If the value of α is equal to 1, there is uniform distribution of topics in the articles. If the value of α is less than 1, there is good separation of topics and the documents are dispersed toward the corners. If the value of α is greater than 1, all the documents are clustered in the center and there is no good separation of topics. The contents of most articles tend to follow the α<1 distribution. The next question is, how about words? We repeat the same notion for the words. In Figure 7-15, we used four words: league, planet, school, and goal. We place the words in the corners and points as the topics.

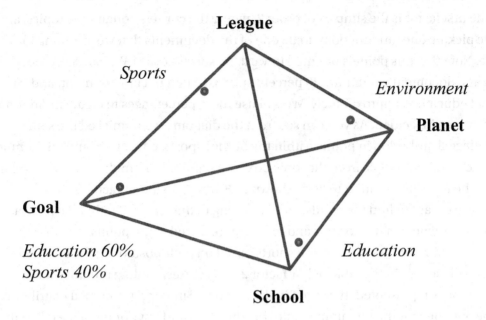

Figure 7-15. *Topics of words*

Let us now put these into the equation perspective and demonstrate the generation of a single document from this LDA. We will break down the formula into four factors for the purpose of this demonstration, as shown in Figure 7-16.

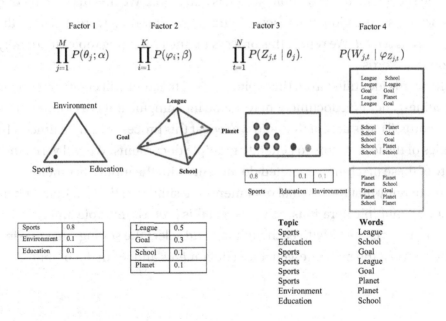

Figure 7-16. *Formula breakdown*

The first factor is the simplex of order 2, where the corners contain the topics, and we are picking one random point to be one of the documents that we are going to create. Note that this point is oriented toward the sports corner. We can represent this point as a document containing 80 percent sports, 10 percent environment, and 10 percent education, approximately. We will use these percentages to create multinomial distributions in factor 3. As you can see from the diagram, we mimicked the same percentage distribution in points within the box for sports, education, and environment. If we pick a random topic from the topic box with replacement multiple times, we can expect the topic distribution to be as shown in Figure 7-16 under factor 3.

Now we want to find the words corresponding to the topics. For this, we pick factor 2, which has words in the corners and topics represented at the points. We quantify the percentage of words associated with sports based on their position in the tetrahedron. This is represented in the table below factor 2. This corresponding multinomial distribution is represented in factor 4 in the first box. Similar multinomial distributions can be obtained for the remaining topics of education and environment based on the factor 2 Dirichlet distributions of education and environment.

Now, we go back to factor 3, topic distributions. We select the first topic, which is sports. We look up the corresponding multinomial distribution of words from factor 4 and select a random word from the first box, which happens to be *league*. We then move to the second instance of topic, which is education. We then use the second box, which represents education, from factor 4 and select a random word. Let's say this random word is *school*. We repeat this process for the remaining topics to identify the corresponding words.

All the words identified from the topics stacked together will create our document. This machine-generated document may not be meaningful and is a combination of different words. That is acceptable, and we repeat this process multiple times, changing the settings of the four factors to generate multiple documents. We will verify the similarity of these machine-generated documents with the input documents. There is a low probability that the original documents are similar to the machine-generated documents in multiple iterations. As we repeat this process multiple times, we observe how the similarity changes with each iteration and select the settings that maximize the similarity between the machine-generated documents and input documents.

LDA Implementation

You may now wonder how you are to know the number of topics when you are presented with a corpus of documents. The answer is through hyperparameter settings. In LDA, the number of topics is also a hyperparameter. Results vary according to the number of topics selected. When we present a corpus of documents to the computer it wouldn't understand any interpretation as humans do. But computers are good at recognizing the frequency of word occurrences in a document. We can also verify how these words are occurring in different documents across the corpus. Using this information, we can tag words to topics. There are two major properties that drive this process, as follows:

- Articles contain a single topic as possible

- Words are associated with a single topic as possible

If we break down the preceding properties, taking an example of four documents as shown here in Figure 7-17, we will find documents that contain a mix of words; usually there is one dominant topic that exists in the documents. We can quantitatively determine the number of occurrences of words and then tag them to different topics accordingly. Say we have chosen three topics; we will classify these four documents into topic 1, topic 2, and topic 3.

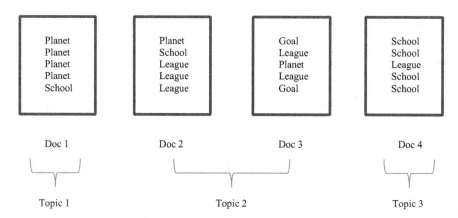

Figure 7-17. *Example corpus*

The second property is demonstrated in the word occurrences by topic. Although there are multiple instances where one word is associated with multiple topics, each word is strongly oriented toward a single topic. For simplicity, assume the majority drives the word assignment to the topic. Then we will have the data shown in Table 7-11.

Table 7-11. *Topic Assignment*

Word	Topic
Planet	Topic 1
Goal	Topic 2
League	Topic 2
School	Topic 3

If there are some commonly occurring words, such as *the, is,* and *any,* they are eliminated using the natural language processing technique of *term frequency – inverse document frequency* (tf–idf). This is a technique used to identify the most important words in a corpus. The tf–idf value increases proportionally to the number of times a word appears in the document and is offset by the number of documents in the corpus that contain the word, which helps to adjust for the fact that some words appear more frequently in general.

We can condense the problem statement as classifying all the words into the given/ chosen number of topics, retaining a single topic for each document if possible and each word into a single topic if possible. This is done through Gibbs sampling. What we are trying to do here is figure out the latent variables that describe what's going on in a topic model. We are assuming a collection of latent variables that completely describe how we are going from topics to allocations to assignments to generate our complete document.

Gibbs sampling allows us to organize all the words one by one with the right topics. This is done in a couple of steps. The first step involves a word search within the document, and the second step involves a word search across documents. Using Figure 7-17, from document 1 let us pick the first word, *planet.* We will not assign this word to any topic, but we count how the other words in the document are assigned to different topics. We have three planets and one school word. These three planet words are assigned to topic 1, and *school* is assigned to topic 3 based on Table 7-11. Now we will look at how the word *planet* is occurring across different topics (Table 7-12).

Table 7-12. *Topic Occurrences*

Topic	Topic Word assignments	Occurrences	Estimates
Topic 1	How much of Topic 1 is in Doc 1	3	$2 + \alpha$
	How much of "Planet" is in topic 1	4	$4 + \beta$
Topic 2	How much of Topic 2 is in Doc 1	0	$0 + \alpha$
	How much of "Planet" is in topic 2	2	$2 + \beta$
Topic 3	How much of Topic 3 is in Doc 1	1	$1 + \alpha$
	How much of "Planet" is in topic 3	0	$0 +$

Using the occurrences, we can calculate the odds of a word's belonging to a topic. If we just use the occurrences alone for each topic, we may run into the problem of estimating 0, giving no chance to other topics for a selected word. To overcome this we add small adjustment terms α and β. If you recall the equation from Figure 7-16, α and β are Dirichlet distributions. We repeat this for all the words in each document multiple times until we maximize the two properties we discussed earlier in the section. A key takeaway from Gibbs sampling is that we are able to reclassify words in a way that words are associated with a single topic if possible. As this process is done iteratively, the computer will be able to help you identify the topics in each document and give you a list of words that define the topic. As a final step, we will have to label these topics, observing the keywords.

We will take a real-world example to understand how LDA can be implemented using PySpark. For this purpose, we will use a dataset from Kaggle (`https://www.kaggle.com/PromptCloudHQ/amazon-echo-dot-2-reviews-dataset`). This dataset contains smart speaker Echo Dot reviews. It contains the following fields:

- Page URL

- Title

- Review Text

- Device Color

- User Verified

- Review Date

- Review Useful Count

- Configuration

- Rating

- Declaration Text (Example: Vine Voice, Top 100 reviewer, etc.)

We will be using the column *Review Text* for our modeling exercise.

```
# Import Sparksession
from pyspark.sql import SparkSession
spark=SparkSession.builder.appName("LDA").getOrCreate()

# Print PySpark and Python versions
```

```python
import sys
print('Python version: '+sys.version)
print('Spark version: '+spark.version)

# Read data
file_location = "lda_data.csv"
file_type = "csv"
infer_schema = "false"
first_row_is_header = "true"

df = spark.read.format(file_type)\
.option("inferSchema", infer_schema)\
.option("header", first_row_is_header)\
.load(file_location)

# Print metadata
df.printSchema()

#  Count data
df.count()
print('The total number of records in the credit card dataset are '+str(df.
count()))

%%bash
pip install nltk

# Import appropriate libraries
from pyspark.sql.types import *
from pyspark.mllib.linalg import Vector, Vectors
from pyspark.ml.feature import CountVectorizer , IDF
from pyspark.mllib.clustering import LDA, LDAModel
from pyspark.mllib.linalg import Vectors as MLlibVectors

import re
import nltk
nltk.download('stopwords')
from nltk.corpus import stopwords
```

In the preceding steps, we are reading the data, importing the necessary libraries, and exploring the metadata associated with the data.

```
reviews = df.rdd.map(lambda x : x['Review Text']).filter(lambda x: x is not
None)
StopWords = stopwords.words("english")
tokens = reviews.map(lambda document: document.strip().lower())\
    .map( lambda document: re.split(" ", document)) \
    .map( lambda word: [x for x in word if x.isalpha()]) \
    .map( lambda word: [x for x in word if len(x) > 3] )\
    .map( lambda word: [x for x in word if x not in StopWords]).
    zipWithIndex()
```

In the preceding steps we are changing the DataFrame to RDD and filtering out the empty reviews. Then we are loading the stopwords from the *ntlk* library and using multiple lambda functions to strip extra spaces, change to lowercase, check alphanumerics, and remove any words less than three letters long, including stopwords. We are finally adding an index to each row to use them as identifiers later in the process.

```
# Convert the rdd to DataFrame
df_txts = spark.createDataFrame(tokens, ['list_of_words','index'])

# TF
cv = CountVectorizer(inputCol="list_of_words", outputCol="raw_features",
vocabSize=5000, minDF=10)
cvmodel = cv.fit(df_txts)
result_cv = cvmodel.transform(df_txts)

# IDF
idf = IDF(inputCol="raw_features", outputCol="features")
idfModel = idf.fit(result_cv)
result_tfidf = idfModel.transform(result_cv)
```

In the preceding steps we are creating term frequency and inverse document frequency using the built-in functions. CountVectorizer helps convert the collection of text documents to vectors of token counts. IDF reweights features to lower values that appear frequently in a corpus.

We are using this data, which contains the words with high importance from each document, in the LDA model. We then are printing out the top five words from each topic to understand the topic. Here we used ten topics, but is this the best? No, we will have to try out different combinations to identify the right number. Ten is used here as an example. You can iterate the following over different combinations and observe which number gives you a meaningful topic collection.

```python
num_topics = 10
max_iterations = 100
lda_model=LDA.train(result_tfidf.select("index", "features").rdd.
mapValues(MLlibVectors.fromML).map(list),k = num_topics, maxIterations =
max_iterations)
```

```python
wordNumbers = 5
data_topics=lda_model.describeTopics(maxTermsPerTopic = wordNumbers)
vocabArray = cvmodel.vocabulary
topicIndices = spark.sparkContext.parallelize(data_tp)
def topic_render(topic):
    terms = topic[0]
    result = []
    for i in range(wordNumbers):
        term = vocabArray[terms[i]]
        result.append(term)
    return result
```

```python
topics_final = topicIndices.map(lambda topic: topic_render(topic)).
collect()
for topic in range(len(topics_final)):
    print ("Topic" + str(topic) + ":")
    for term in topics_final[topic]:
        print (term)
    print ('\n')
```

Output:

```
Topic0:
works
good
speaker
sound
better

Topic1:
know
time
questions
playing
alexa

Topic2:
amazon
another
first
purchased
back

Topic3:
would
device
work
could
give

Topic4:
well
need
thing
pretty
loves

Topic5:
```

love
great
bought
little
much

Topic6:
home
weather
smart
voice
things

Topic7:
really
getting
without
wish
enjoy

Topic8:
music
echo
play
every
want

Topic9:
product
easy
still
alexa
using

Collaborative Filtering

If you have used or have been presented with any recommendation engines, the chances are highly likely that collaborative filtering was used under the hood. One of the advantages of collaborative filtering is that it doesn't need an extensive load of customer data except for historical preference on a set of items. One of the inherent assumptions is that future preferences are consistent with past preferences. There are two major categories of user preferences, Explicit and Implicit ratings. As the name suggests, Explicit ratings are given by the user on an agreed service provider scale (e.g., 1 to 10 for IMDB). This presents a great opportunity, as we know how strongly a user likes or dislikes a product/content. Implicit ratings, on other hand, are measured through engagement activities (e.g., how many times did a customer listen to a particular track, purchases, viewing, etc.).

Data in the real world is usually bimodal. It is created as a joint interaction between two types of entities. For example, a user rating a movie is affected by both user and movie characteristics. If you imagine a dataset with rows containing different users and columns containing their ratings of movies, classical clustering techniques would look at how close the users are in the multi-dimensional space. We often ignore the relationship between the columns. Co-clustering aims to group objects by both rows and columns simultaneously based on the similarity of their pair-wise interactions. Co-clustering is powerful when we have sparse data. Data is sparse in most real-world cases. A viewer can watch a finite number of movies, so as we pick data across users and movies, data will be sparse, and this is one of the best examples of co-clustering. In the collaborative filtering lingo, we identify them as user-based collaborative filtering and item-based collaborative filtering.

User-based Collaborative Filtering

Say we have $a \times b$ matrix of ratings with users u_i, i =1 to n; and items p_j, j=1 to m; if we wanted to predict rating r_{ij} and if the user did not watch item j, we would calculate similarities between user i and all other users (Figure 7-18). We would then select top similar users and take the weighted average of ratings from these top users with similarities as weights.

$$r_{ij=} \frac{\sum_k Similarities\left(u_i, u_k\right) r_{kj}}{number\ of\ ratings}$$

Again, we may have different types of raters ranging from strict to lenient. To avoid the bias, we will make use of each user's average rating of all items when computing and add it back to the target.

$$r_{ij=} \ \overline{r_i} \ + \ \frac{\sum_k Similarities \left(u_i, u_k \right) \left(r_{kj} - \overline{r_k} \right)}{number \ of \ ratings}$$

There are other ways to calculate similarity, such as Pearson correlation and cosine similarity. All these metrics allow us to find the most similar users to your target user and are based on the principle that users are similar when they engage equally or purchase similar items.

Item-based Collaborative Filtering

For item-based collaborative filtering, we consider two items to be similar when they receive similar ratings from the same user. We will then make a prediction on target on an item by calculating the weighted average of the ratings for the most similar items from this user. This approach offers us stability but is not computationally efficient.

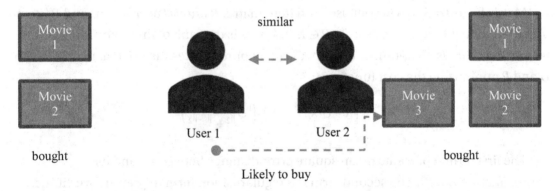

Figure 7-18. *User-based collaborative filtering*

Matrix Factorization

Co-clustering requires the use of matrix factorization. It is a method of decomposing an initial sparse matrix of user- and item-based interactions into low-dimensional matrices with latent features' having lower sparsity. Latent features can be a broad categorization of items (e.g., fresh produce in grocery, genres in movies). Matrix factorizations gives us an estimation of how much a movie fits into a category and how much a user fits into the same category. One of the greatest benefits of this method is even if there is no explicit or implicit rating of common items between two users, we will be able to find the similarity using the latent features that they share (Figure 7-19).

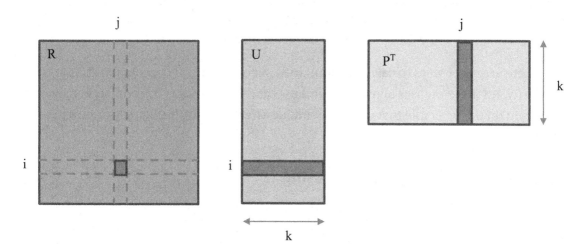

Figure 7-19. *Matrix factorization*

Matrix factorization will split user and item matrix R into U (user based) and P (item based) such that $R \sim U \times P$, $U \in R^{m*k}$; $P \in R^{n*k}$. k here is the rank of the factorization. Now we can treat this as an optimization process that approximates the original matrix R with U and P minimizing the cost function J:

$$J = \left\| R - U \times P^T \right\|_2 + \lambda \left(\left\| U \right\|_2 + \left\| P \right\|_2 \right)$$

The first term represents mean square error distance between R and its approximation $U \times P^T$. The second term is a regularization term to prevent overfitting.

Optimization Using Alternating Least Squares (ALS)

ALS is used to minimize the loss function by changing half of the parameters. By fixing half of the parameters you recompute the other half and repeat and there is no gradient in the optimization step. It is an efficient solution to matrix factorization for implicit feedback. The costs of gradient computation can be extremely hard as we assume unobserved entries to be 0. You will gain deeper insight as we discus more about the cost function.

Based on the cost function, we are trying to learn two types of variables, U and P, which are represented as $U \times P^T$. The actual cost of the function is the following sum:

$$\left\| R - U \times P^T \right\|_2 = \sum_{i,j} \left(R_{i,j} - u_i \times p_j \right)$$

Plus the regularization term. Since U's and V's are unknown variables, it makes these cost functions non-convex. If we fix P and optimize for U, we essentially can reduce the problem to linear regression. This is a massive win, and the process can be parallelized. ALS is a two-step iterative process. In each iteration, P is fixed and U is solved. In the subsequent step, U is fixed and P is solved. Since ordinary least squares guarantees the minimal mean squared error, the cost function can never increase. Alternating between the steps will reduce the cost function till convergence.

Collaborative filtering is great for recommendations when the least amount of data is available. But you have to be aware of its limitations. The latent features that we generate as part of matrix decomposition are not interpretable. We can also run into the issue of a cold start. When a new item doesn't have a sizable number of users, the model will fail in making a recommendation of the new item and can fall prey to popularity bias.

We will now demonstrate the implementation of collaborative filtering by using a made-up dataset containing three columns: *userid*, *movieid*, and *ratings* of multiple users and movies. All the code will be provided in separate Jupyter notebooks for easy access.

```
# Import Sparksession
from pyspark.sql import SparkSession
spark=SparkSession.builder.appName("CF").getOrCreate()

# Print PySpark and Python versions
import sys
print('Python version: '+sys.version)
print('Spark version: '+spark.version)

# Read data
file_location = "cf_data.csv"
file_type = "csv"
infer_schema = "false"
first_row_is_header = "true"
```

```
df = spark.read.format(file_type)\
.option("inferSchema", infer_schema)\
.option("header", first_row_is_header)\
.load(file_location)

# Print metadata
df.printSchema()

#  Count data
df.count()
print('The total number of records in the credit card dataset are '+str(df.
count()))

# Import appropriate libraries
from pyspark.sql.types import *
import pyspark.sql.functions as sql_fun
from pyspark.ml.recommendation import ALS, ALSModel
from pyspark.mllib.evaluation import RegressionMetrics, RankingMetrics
from pyspark.ml.evaluation import RegressionEvaluator
import re

# Casting variables
int_vars=['userId','movieId']
for column in int_vars:
    df=df.withColumn(column,df[column].cast(IntegerType()))
float_vars=['rating']
for column in float_vars:
    df=df.withColumn(column,df[column].cast(FloatType()))

(training, test) = df.randomSplit([0.8, 0.2])
```

Observe we are using ALS optimization to achieve matrix factorization. Parameters of the ALS model in PySpark are as follows:

- NumBlocks is the number of blocks the users and items will be partitioned into in order to parallelize computation.

- rank is the number of latent features in the model.

- `maxIter` is the maximum number of iterations to run.

- `regParam` specifies the regularization parameter in ALS.

- `implicitPrefs` specifies whether to use the explicit feedback ALS variant or one adapted for implicit feedback data (defaults to false, which means using explicit feedback).

- `alpha` is a parameter applicable to the implicit feedback variant of ALS that governs the baseline confidence in preference observations (defaults to 1.0).

We set `coldStartStrategy` to drop to eliminate empty evaluation metrics.

```
als = ALS(rank=15,maxIter=2, regParam=0.01,
          userCol="userId", itemCol="movieId", ratingCol="rating",
          coldStartStrategy="drop",
          implicitPrefs=False)
model = als.fit(training)

# Evaluate the model by computing the RMSE on the test data
predictions = model.transform(test)
evaluator = RegressionEvaluator(metricName="rmse", labelCol="rating",
                                predictionCol="prediction")

rmse = evaluator.evaluate(predictions)
print("Root-mean-square error = " + str(rmse))

# Generate top 10 movie recommendations for each user
userRecs = model.recommendForAllUsers(10)
userRecs.count()

# Generate top 10 user recommendations for each movie
movieRecs = model.recommendForAllItems(10)
movieRecs.count()

userRecs_df = userRecs.toPandas()
print(userRecs_df.shape)

movieRecs_df = movieRecs.toPandas()
print(movieRecs_df.shape)
```

```
userRecs_df.head()
```

Output:

	userId	recommendations
0	190174	[(1338, 9.24465560913086), (72028, 9.121529579...
1	190227	[(3812, 7.743837356567383), (2662, 7.072795867...
2	190387	[(3943, 9.847469329833984), (2092, 9.769483566...
3	190540	[(361, 9.0469970703125), (2489, 8.711124420166...
4	190348	[(3912, 8.182646751403809), (995, 8.0464324951...

EXERCISE : COLLABORATIVE FILTERING

Question 1: Change the number of latent features in the ALS settings and observe how the mean squared error changes.

Summary

- We got an overview of unsupervised learning techniques.

- We now know the differences between hierarchical and non-hierarchical clustering methods and Silhouette coefficients to determine the right number of clusters.

- We explored how latent Dirichlet allocation works and implemented it.

- We also reviewed the implementation of collaborative filtering using alternating least squares.

Great job! You are now familiar with some of the key concepts that will be useful in unsupervised learning and unstructured data. We will take you through a journey of how you can structure your projects and create automated machine learning flows in the next chapter.

CHAPTER 8

Machine Learning Flow and Automated Pipelines

Putting a model into production is one of the most challenging tasks in the data science world. It is one of those last-mile problems that persists in many organizations. Although there are many tools for managing workflows, as the organization matures its needs change, and managing existing models can become a herculean task. When we take a step back and analyze why it is so challenging, we can see that it is because of the structure that exists in most organizations. There is an engineering team that maintains the production platform. There is a gap between the data science toolset and the production platforms. Some of the data science work can be developed in Jupyter Notebook, with little consideration given to the cloud environment. Some of the data flows are created locally with limited scaling. Such applications tend to falter with large amounts of data. Best practices that exist in the software development cycle don't stick well with the machine learning lifecycle because of the variety of tasks involved. The standard is mainly defined by the data science team in the organization. Also, rapid developments in the field are leaving vacuums with respect to the management and deployment of models.

The goal of this chapter is to explore the tools available for managing and deploying data science pipelines. We will walk you through some of the tools available in the industry and demonstrate with an example how to construct an automated pipeline. We will use all the concepts that we have learned so far to construct these pipelines. By the end of this chapter, you will have learned how to build the machine learning blocks and automate them in coherence with custom data preprocessing.

© Ramcharan Kakarla, Sundar Krishnan and Sridhar Alla 2021
R. Kakarla et al., *Applied Data Science Using PySpark*, https://doi.org/10.1007/978-1-4842-6500-0_8

We will cover the following topics in this chapter:

- MLflow. We will introduce MLflow as well as its components and advantages, followed by a demonstration. In the first half, our focus will be on the installation and introduction of concepts. In the second half, we will focus on implementing a model in MLflow.

- Automated machine learning pipelines. We will introduce the concept of designing and implementing automated ML frameworks in PySpark. The first half of the section is heavily focused on implementation, and the second half focuses on the outputs generated from the pipeline.

MLflow

With the knowledge obtained from previous chapters, we know how to build individual supervised and unsupervised models. Often times we would try multiple iterations before picking the best model. How do we keep track of all the experiments and the hyperparameters we used? How can we reproduce a particular result after multiple experiments? How do we move a model into production when there are multiple deployment tools and environments? Is there a way to better manage the machine learning workflow and share work with the wider community? These are some of the questions that you will have after you are comfortable building models in PySpark. Databricks has built a tool named MLflow that can gracefully handle some of the preceding questions while managing the machine learning workflow. MLflow is designed to be open interface and open source. People often tend to go back and forth between experimentation and adding new features to the model. MLflow handles such experimentation with ease. It is built around REST APIs that can be consumed easily by multiple tools. This framework also makes it easy to add existing machine learning blocks of code. Its open source nature makes it makes it easy to share the workflows and models across different teams. In short, MLflow offers a framework to manage the machine learning lifecycle, including experimentation, reproducibility, deployment, and central model registry.

MLflow components are shown in Figure 8-1 and described in the following list.

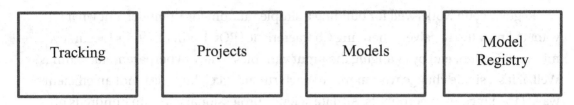

Figure 8-1. *MLflow components*

- **Tracking:** Aids in recording, logging code, configurations, data, and results. It also gives a mechanism to query experiments.

- **Projects:** This component of MLflow helps in bundling the machine learning code in a format for easy replication on multiple platforms.

- **Models:** This assists in deploying machine learning code in diverse environments.

- **Model Registry:** This component acts like a repository in which to store, manage, and search models.

This framework opens the tunnel for data engineers and data scientists to collaborate and efficiently build data pipelines. It also provides the ability to consume and write data from and to many different systems. It opens up a common framework for working with both structured and unstructured data in data lakes as well as on batch streaming platforms.

This may look abstract for first-time users. We will walk you through an example of how these components can be useful and implemented. For this purpose of illustration, let us take the same bank dataset we used in Chapter 6.

MLflow Code Setup and Installation

In this section, we will walk through the code changes necessary to support MLflow, along with its installation. We will be using a PySpark Docker version for all demonstrations. There are two main steps to get MLflow up and running, as follows:

- Code/script changes

- Docker side changes and MLflow server installation

Code changes are minimal. We have to add a few MLflow functions to existing machine learning code to accommodate MLflow.

Regular code works well for building a simple random forest model, but what if I want to know the Receiver Operating Characteristic (ROC)/other metrics (accuracy, misclassification, etc) by changing the input variables or any hyperparameter settings? Well, if it's a single change we can record each run in Excel, but this is not an efficient way of tracking our experiments. As data science professionals, we run hundreds of experiments, especially in the model fine-tuning phase. Is there a better way of capturing and annotating these results and logs? Yes, we can do it via the MLflow tracking feature. Now, let's rewrite a simple random forest model code, adding MLflow components. Changes in the code are highlighted in bold.

Note   We will save the following code as .py and run via Docker after installing MLflow.

```
# Import libraries
import pyspark
from pyspark.sql import SparkSession
import mlflow
import mlflow.spark
import sys
import time
from pyspark.ml.classification import RandomForestClassifier
from pyspark.ml.evaluation import BinaryClassificationEvaluator
```

Observe we have imported the libraries related to MLflow that allow the user to use PySpark. Note MLflow supports a variety of languages and packages, including Spark and Python, with extended support for packages like scikitlearn and tensorflow.

```
spark = SparkSession.builder.appName("mlflow_example").getOrCreate()

filename = "/home/jovyan/work/bank-full.csv"
target_variable_name = "y"
from pyspark.sql import functions as F
df = spark.read.csv(filename, header=True, inferSchema=True, sep=';')
df = df.withColumn('label', F.when(F.col("y") == 'yes', 1).otherwise(0))
df = df.drop('y')
train, test = df.randomSplit([0.7, 0.3], seed=12345)
```

```python
for k, v in df.dtypes:
    if v not in ['string']:
        print(k)

df = df.select(['age', 'balance', 'day', 'duration', 'campaign', 'pdays',
'previous', 'label'])

from pyspark.ml.feature import VectorAssembler
from pyspark.ml import Pipeline

def assemble_vectors(df, features_list, target_variable_name):
    stages = []
    #assemble vectors
    assembler = VectorAssembler(inputCols=features_list,
    outputCol='features')
    stages = [assembler]
    #select all the columns + target + newly created 'features' column
    selectedCols = [target_variable_name, 'features']
    #use pipeline to process sequentially
    pipeline = Pipeline(stages=stages)
    #assembler model
    assembleModel = pipeline.fit(df)
    #apply assembler model on data
    df = assembleModel.transform(df).select(selectedCols)

    return df

#exclude target variable and select all other feature vectors
features_list = df.columns
#features_list = char_vars #this option is used only for ChiSqselector
features_list.remove('label')

# apply the function on our DataFrame
assembled_train_df = assemble_vectors(train, features_list, 'label')
assembled_test_df = assemble_vectors(test, features_list, 'label')

print(sys.argv[1])
print(sys.argv[2])
```

These are the system arguments we intend to change in each experiment. We will pass this as system arguments when we execute this script. Here we used maxBins and maxDepth. You are free to include any other variable changes you want to log across the script.

```
from pyspark.ml.classification import RandomForestClassifier
from pyspark.ml.evaluation import BinaryClassificationEvaluator

maxBinsVal = float(sys.argv[1]) if len(sys.argv) > 3 else 20
maxDepthVal = float(sys.argv[2]) if len(sys.argv) > 3 else 3
```

In the following step, we are initializing the machine learning code part with MLflow since our tracking variables of interest are within this scope. You can modify the scope as intended depending upon the use case.

```
with mlflow.start_run():
    stages_tree=[]
    classifier = RandomForestClassifier(labelCol = 'label',featuresCol =
    'features',maxBins=maxBinsVal, maxDepth=maxDepthVal)
    stages_tree += [classifier]
    pipeline_tree=Pipeline(stages=stages_tree)
    print('Running RFModel')
    RFmodel = pipeline_tree.fit(assembled_train_df)
    print('Completed training RFModel')
    predictions = RFmodel.transform(assembled_test_df)
    evaluator = BinaryClassificationEvaluator()
    print("Test Area Under ROC: " + str(evaluator.evaluate(predictions,
    {evaluator.metricName: "areaUnderROC"})))

    mlflow.log_param("maxBins", maxBinsVal)
    mlflow.log_param("maxDepth", maxDepthVal)
    mlflow.log_metric("ROC", evaluator.evaluate(predictions, {evaluator.
    metricName: "areaUnderROC"}))
    mlflow.spark.log_model(RFmodel,"spark-model")
```

In the preceding snippet of code, we added log_param and log_metric to capture the pieces of information we want to keep tabs on. Also note we are logging the model using the mlflow.spark.log_model function, which helps in saving the model on the MLflow backend. This is an optional statement, but it is handy if you want to register the model from MLflow. With minimal changes in code, we are able to accommodate an existing model with MLflow components. Why is it so important? Tracking the change in parameters and metrics can become challenging with hundreds of experiments staged over multiple days. We will need to save the preceding code as a Python file. We will be using a Spark submit to execute this PySpark code. In this illustration, we saved it as *Chapter9_mlflow_example.py*.

Moving on to part two of MLflow setup, it is time to initiate the Docker container using the following command. Recall this is the same PySpark Docker image we demonstrated in Chapter 1. We also have to save the preceding Python file in the same local path that we are mapping to the Docker container. The following Docker initiation code and its local mapping works for Mac; for Windows, please change your path appropriately.

```
docker run -it -p 5000:5000 -v /Users/ramcharankakarla/demo_data/:/home/
jovyan/work/ jupyter/pyspark-notebook:latest bash
```

Note   MLflow by default uses port 5000.

As we enter into Docker, we like to change the path to work (cd /home/jovyan/ work/) since all our local information, including the Python file *Chapter9_mlflow_ example.py,* is mapped there. The following two commands help you to change directory and see the contents available on the work directory in Docker.

```
cd work
ls
```

Make sure you see the Python file in this path. Often times the Docker image may not contain the MLflow. It is a good practice to verify if the MLflow package is available in the image by using pip freeze|grep mlflow. If MLflow is not available, add the package using the following command:

```
pip install mlflow
```

The next step is to initialize the MLflow server. This server is the backend to the user interface and captures all the information pertaining to the experiments. It can be initiated using the following command. Note that we are running this command in the background.

```
mlflow server --backend-store-uri sqlite:///mlflow.db --default-artifact-
root /home/jovyan/work --host 0.0.0.0 --port 5000 &
```

Note the parameters here contain backend-store-uri and default-artifact-root. Backend-store-uri is mandatory if we want to register or log any models. We used a default sqlite MLflow data base (db) here for demonstration. This can be any other backend store, including file-based and database-backend stores. This powers the querying capability of MLflow. Artifact root is the location of the artifact store and is suitable for large data. This is where all the artifacts pertaining to the experiments reside. By default the location of --default-artifact-root is set to ./mlruns. You can override this setting by giving an appropriate location. Different artifact stores are supported by MLflow including, the following:

- HDFS

- NFS

- FTP server

- SFTP server

- Google Cloud storage (GCS)

- Azure Blob storage

- Amazon S3

It is important to specify the artifact URI when creating an experiment, or else the client and server will refer to different physical locations; i.e., the same path on different disk locations (Table 8-1).

Table 8-1. *Storage Types*

Storage Type	URI Format
FTP	ftp://user:pass@host/path/to/directory
SFTP	sftp://user@host/path/to/directory
NFS	/mnt/nfs
HDFS	hdfs://<host>:<port>/<path>
Azure	wasbs://<container>@<storage-account>.blob.core.windows.net/<path>
Amazon S3	s3://<bucket>/<path>

There are additional settings needing to be authenticated for a few storage types. For S3, credentials can be obtained from environment variables `AWS_ACCESS_KEY_ID` and `AWS_SECRET_ACCESS_KEY` or from the IAM profile.

MLflow User Interface Demonstration

In this section, we will walk through the MLflow user interface with an example we discussed in the previous section. The following are running the script we saved earlier with three different settings:

```
MLFLOW_TRACKING_URI="http://0.0.0.0:5000" spark-submit --master local[*] /
home/jovyan/work/Chapter9_mlflow_example.py 16 3  --spark_autolog True
```

```
MLFLOW_TRACKING_URI="http://0.0.0.0:5000" spark-submit --master local[*] /
home/jovyan/work/Chapter9_mlflow_example.py 16 5  --spark_autolog True
```

```
MLFLOW_TRACKING_URI="http://0.0.0.0:5000" spark-submit --master local[*] /
home/jovyan/work/Chapter9_mlflow_example.py 32 5  --spark_autolog True
```

Once we finish running the preceding three commands, we can proceed with initiating the MLflow user interface. On any web browser, navigate to `http://0.0.0.0:5000`. You will see the MLflow user interface shown in Figure 8-2.

As you can see, the user interface has all the information formatted into experiments and models. We can make any annotations or notes using the notes options pertaining to each of the runs within each experiment. This framework gives us the ability to manage and run multiple experiments with multiple runs. We also have an option to filter the runs based on parameter settings. This is extremely useful in a machine learning lifecycle because, unlike the software development lifecycle, we tend to iterate between old and newer versions based on the stability and accuracy of the models.

Figure 8-2. *MLflow UI*

There is also detailed row-level information on each run. We can drill down and compare this information across many experiments. For our illustration, we have only used a single metric, but in a real-world scenario, we may also want to capture accuracy, misclassification, lift, and KS statistics for each run. We could then compare and sort each iteration based on a metric that is acceptable based on business need.

When we select a couple of iterations from the prior executions, observe the *Compare* button is active; it is greyed out in Figure 8-2. When we click on *Compare* it will compare these two iterations and also give prepared graphs on how metrics are varying across different runs (Figure 8-3). In the preceding code we have explicitly set the log metrics. From version 3 of Spark there is an option for automatic logging.

Figure 8-3. *MLflow UI comparison of runs*

Observe the window in Figure 8-4; MLflow autogenerates the runids. We can compare multiple runs in an experiment using this feature. When we have enough data points, we can visually pick the model runs that are giving the best performance. We can also observe how the model performance can vary by changing the parameters.

Figure 8-4. *MLflow UI comparison window*

If we go back to the homepage in the MLflow UI and click on any run, it will give us more information about the model run, as shown in Figure 8-5. This Run window also gives us the handy feature of registering the model. Registering a model will make it available for scoring. There are multiple other features that will open up upon registering the model.

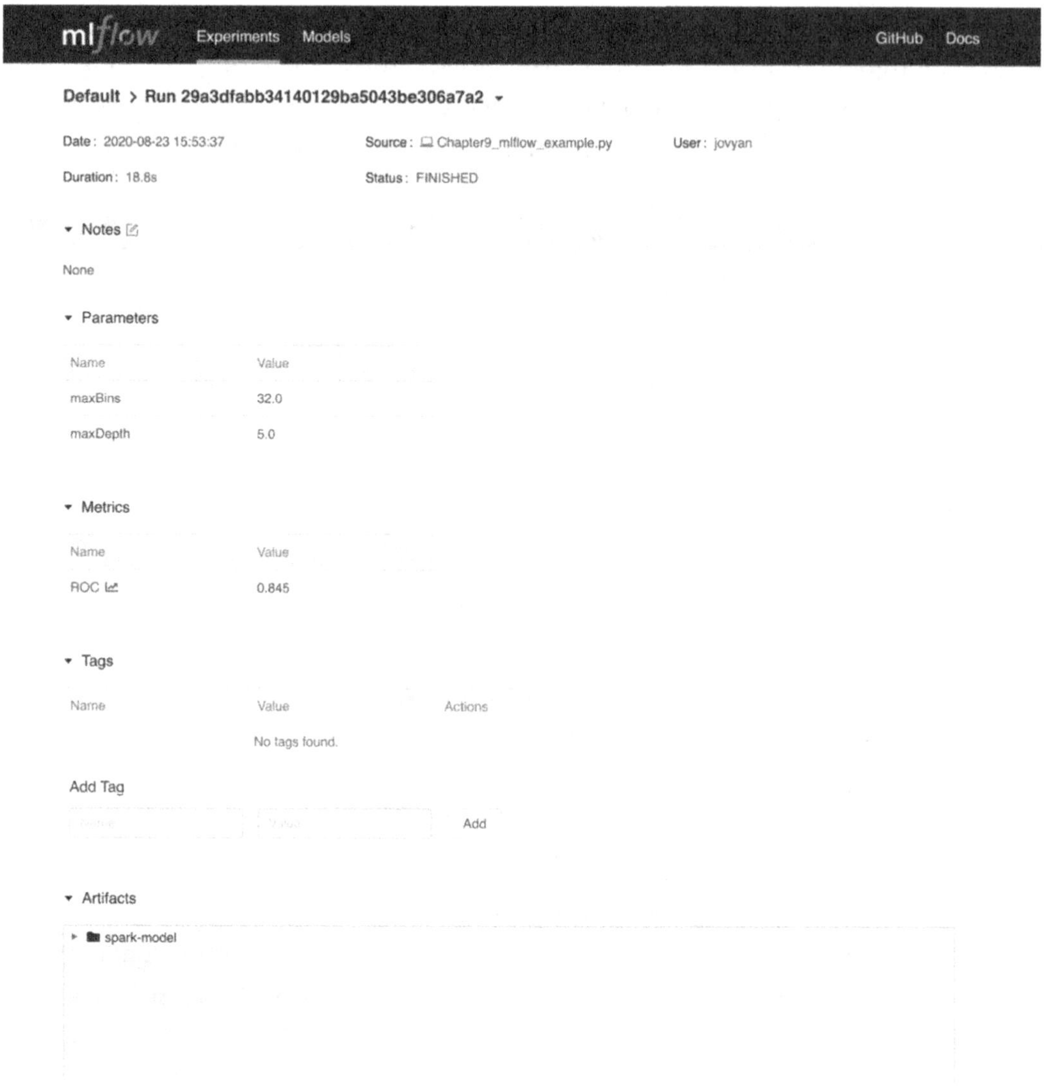

Figure 8-5. *MLflow UI Run window*

Clicking the *spark-model* file in the Artifacts section will open up the next window. This will contain all the metadata information of the model required for running it, as shown in Figure 8-6.

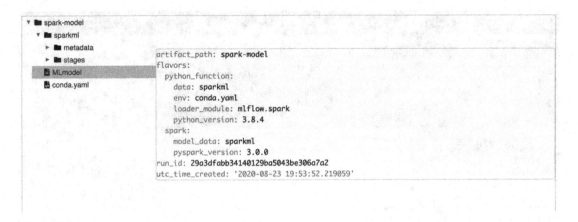

Figure 8-6. *MLflow UI model information*

If you click the root folder it will give an option to register the model in the metadata, as shown in Figure 8-7.

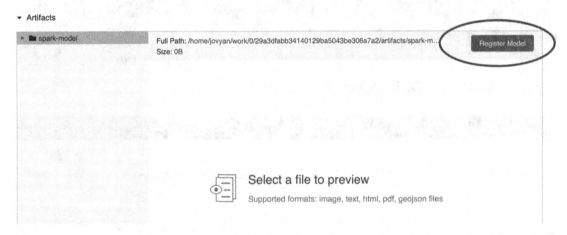

Figure 8-7. *MLflow UI model registration*

Adding the model as shown in Figure 8-8 pushes the model into the Models tab. You can verify this by navigating to the Models tab at the top (Figure 8-9).

Figure 8-8. *MLflow UI Registration window*

Figure 8-9. *MLflow Models tab*

On clicking the model name, you will navigate to a new window with detailed model information. The new window will also give options to push the model into three different environments, as shown in Figure 8-10.

- Staging

- Production

- Archived

This helps in effectively managing the model lifecycle by moving models to different environments as we transition from model development to the end of the model lifecycle.

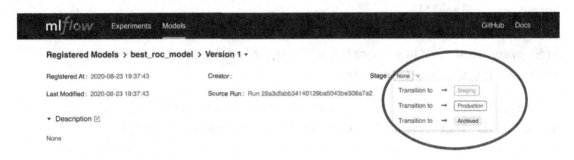

Figure 8-10. *MLflow model environments*

Well, all this is great, but how can we score new data from these pipelines? When we click on either the run or the model information, we get the full path of the model metadata information, represented in Figure 8-7. We can use this metadata to score any new data. There are multiple ways this can be achieved. Let us write a simple code to prepare data for scoring, and we will use the best_roc_model shown in Figure 8-10. We will save the following code as a Python file and run it via *spark-submit* in Docker.

```
#Importing necessary libraries
import mlflow
import mlflow.spark
import sys
from pyspark.sql import SparkSession

spark = SparkSession.builder.appName("mlflow_predict").getOrCreate()

filename = "/home/jovyan/work/bank-full.csv"
target_variable_name = "y"
from pyspark.sql import functions as F
df = spark.read.csv(filename, header=True, inferSchema=True, sep=';')
df = df.withColumn('label', F.when(F.col("y") == 'yes', 1).otherwise(0))
df = df.drop('y')
train, test = df.randomSplit([0.7, 0.3], seed=12345)

for k, v in df.dtypes:
```

```
    if v not in ['string']:
        print(k)

df = df.select(['age', 'balance', 'day', 'duration', 'campaign', 'pdays',
'previous', 'label'])

from pyspark.ml.feature import VectorAssembler
from pyspark.ml import Pipeline

# assemble individual columns to one column - 'features'
def assemble_vectors(df, features_list, target_variable_name):
    stages = []
    #assemble vectors
    assembler = VectorAssembler(inputCols=features_list,
    outputCol='features')
    stages = [assembler]
    #select all the columns + target + newly created 'features' column
    selectedCols = [target_variable_name, 'features']
    #use pipeline to process sequentially
    pipeline = Pipeline(stages=stages)
    #assembler model
    assembleModel = pipeline.fit(df)
    #apply assembler model on data
    df = assembleModel.transform(df).select(selectedCols)

    return df


#exclude target variable and select all other feature vectors
features_list = df.columns
#features_list = char_vars #this option is used only for ChiSqselector
features_list.remove('label')

# apply the function on our dataframe

assembled_test_df = assemble_vectors(test, features_list, 'label')

print(sys.argv[1])
```

```
# model information from argument
model_uri=sys.argv[1]
print("model_uri:", model_uri)
model = mlflow.spark.load_model(model_uri)
print("model.type:", type(model))
predictions = model.transform(assembled_test_df)
print("predictions.type:", type(predictions))
predictions.printSchema()
df = predictions.select('rawPrediction','probability', 'label', 'features')
df.show(5, False)
```

This code is a simple pipeline. We also have the flexibility of wrapping the scoring function as a simple udf as well. After saving the code, we need to run the following command in Docker for results:

```
spark-submit --master local[*] Chapter9_predict_spark.py /home/jovyan/work/
0/29a3dfabb34140129ba5043be306a7a2/artifacts/spark-model
```

This will give us the output shown in Figure 8-11.

```
+----------+-----+-------------------------------------------+
|prediction|label|features                                   |
+----------+-----+-------------------------------------------+
|0.0       |1    |[19.0,134.0,27.0,271.0,2.0,-1.0,0.0]       |
|0.0       |0    |[19.0,0.0,11.0,123.0,3.0,-1.0,0.0]         |
|0.0       |0    |[19.0,1169.0,6.0,463.0,18.0,-1.0,0.0]      |
|0.0       |1    |[20.0,423.0,16.0,498.0,1.0,-1.0,0.0]       |
|0.0       |0    |[20.0,-103.0,13.0,180.0,1.0,-1.0,0.0]      |
+----------+-----+-------------------------------------------+
only showing top 5 rows
```

Figure 8-11. *MLflow model scoring output*

All the preceding content and codes are specified in Spark. There are multiple other flavors that MLflow supports. We can serialize the preceding pipeline in the mleap flavor, which is a project to host Spark pipelines without a Spark context for smaller datasets where we don't need any distributed computing. MLflow also has the capability to publish the code to GitHub in the MLflow project format, making it easy for anyone to run the code.

Automated Machine Learning Pipelines

The machine learning lifecycle is an iterative process. We often tend to go back and forth tuning parameters, inputs and data pipelines. This can quickly get cumbersome with the number of data management pipelines. Creating automated pipelines can save a significant amount of time. From all the earlier chapters we have learned data manipulations, algorithms, and modeling techniques. Now, let us put them all together to create an automated PySpark flow that can generate baseline models for quick experimentation.

For this experiment, we will use a churn dataset from Kaggle (`https://www.kaggle.com/shrutimechlearn/churn-modelling#Churn_Modelling.csv`). This banking dataset contains data about attributes of customers and who has churned. Our objective is to identify customers who will churn based on given attributes. Table 8-2 lists the attributes.

Table 8-2. *Metadata*

Column	Description
RowNumber	Identifier
CustomerId	Unique ID for customer
Surname	Customer's last name
Credit score	Credit score of the customer
Geography	The country to which the customer belongs
Gender	Male or female
Age	Age of customer
Tenure	Number of years for which the customer has been with the bank
Balance	Bank balance of the customer
NumOfProducts	Number of bank products the customer is utilizing
HasCrCard	Binary flag for whether the customer holds a credit card with the bank or not
IsActiveMember	Binary flag for whether the customer is an active member with the bank or not
EstimatedSalary	Estimated salary of the customer in dollars
Exited	Binary flag: 1 if the customer closed account with bank and 0 if the customer is retained

Pipeline Requirements and Framework

When designing any pipeline, it is imperative to define what outputs and requirements we expect. First, let us define the outputs that we think are necessary. Since we are tackling a binary target, it would be nice to have the following outputs:

- **KS:** Kolmogorov-Smirnov test is a measure of separation between goods and the bads. Higher is better..

- **ROC:** (ROC) Curve is a way to compare diagnostic tests. It is a plot of the true positive rate against the false positive rate. Higher is better.

- **Accuracy:** (true positives+true negatives)/ (true positives+true negatives+false positives+false negatives). Higher is better.

Now, let's list the requirements for an automated model building tool.

- Module should be able to handle the missing values and categorical values by itself.

- We also prefer the module to take care of variable selection.

- We also want this module to test out multiple algorithms before selecting the final variables.

- Let the module compare different algorithms based on preferred metric and select the champion and challenger models for us.

- It would be handy if it could remove certain variables by prefix or suffix or by variable names. This can power multiple iterations and tweak input variables.

- Module should be able to collect and store all the output metrics and collate them to generate documents for later reference.

- Finally, it would be nice if the module could save the model objects and autogenerate the scoring codes so we can just deploy the selected model.

How are we writing these requirements? Remember we discussed the CRISP–DM framework in Chapter 3 (Figure 8-12)? We are making use of this framework here as our architecture blueprint.

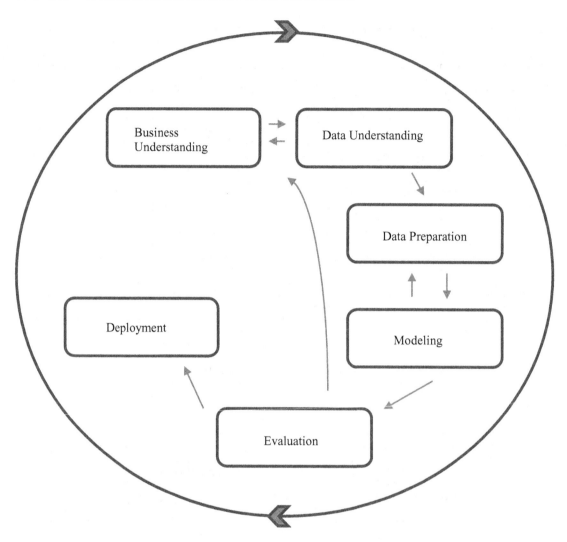

Figure 8-12. *CRISP–DM*

Now, these are a lot of requirements and can quickly become overwhelming if not handled properly. We defined the outputs and requirements, but what will be our expected inputs? Probably a dataset either from a datastore or a flatfile. With this specified information, we can start building the pipelines step-by-step by breaking the code into logical steps. This will be a code-intensive chapter, and you will be able to use this module and create your own baseline automated models. We will break tasks into subtasks in each of the logical steps. These can be broadly divided into the following:

- Data manipulations

- Feature selection

- Model building

- Metrics calculation

- Validation and plot generation

- Model selection

- Score code creation

- Collating results

- Framework to handle all the preceding steps

Data Manipulations

In this section, we will define some commonly used functions for handling the data, such as the following:

- Missing value percentage calculations

- Metadata categorization of input data

- Handling categorical data using label encoders

- Imputing missing values

- Renaming categorical columns

- Combining features and labels

- Data splitting to training, testing, and validation

- Assembling vectors

- Scaling input variables

```
from pyspark.ml.feature import StringIndexer
from pyspark.ml import Pipeline
from pyspark.sql import *
from pyspark.sql.types import *
from pyspark.sql import functions as F
from pyspark.ml.feature import IndexToString
from pyspark.sql.functions import col
```

```
from pyspark.ml.feature import StandardScaler
from pyspark.ml.feature import VectorAssembler

#    1. Missing value calculation

def missing_value_calculation(X, miss_per=0.75):

    missing = X.select([F.count(F.when(F.isnan(c) | F.col(c).isNull(), c)).
    alias(c) for c in X.columns])
    missing_len = X.count()
    final_missing = missing.toPandas().transpose()
    final_missing.reset_index(inplace=True)
    final_missing.rename(columns={0:'missing_count'},inplace=True)
    final_missing['missing_percentage'] = final_missing['missing_count']/
    missing_len
    vars_selected = final_missing['index'][final_missing['missing_
    percentage'] <= miss_per]
    return vars_selected

#    2. Metadata categorization

def identify_variable_type(X):

    l = X.dtypes
    char_vars = []
    num_vars = []
    for i in l:
        if i[1] in ('string'):
            char_vars.append(i[0])
        else:
            num_vars.append(i[0])
    return char_vars, num_vars

#    3. Categorical to Numerical using label encoders

def categorical_to_index(X, char_vars):
    chars = X.select(char_vars)
```

```python
    indexers = [StringIndexer(inputCol=column, outputCol=column+"_
    index",handleInvalid="keep") for column in chars.columns]
    pipeline = Pipeline(stages=indexers)
    char_labels = pipeline.fit(chars)
    X = char_labels.transform(X)
    return X, char_labels
```

```python
#    4. Impute Numerical columns with a specific value. The default is set
to 0.
```

```python
def numerical_imputation(X,num_vars, impute_with=0):
    X = X.fillna(impute_with,subset=num_vars)
    return X
```

```python
#    5. Rename categorical columns
```

```python
def rename_columns(X, char_vars):
    mapping = dict(zip([i+ '_index' for i in char_vars], char_vars))
    X = X.select([col(c).alias(mapping.get(c, c)) for c in X.columns])
    return X
```

```python
#    6. Combining features and labels
```

```python
def join_features_and_target(X, Y):

    X = X.withColumn('id', F.monotonically_increasing_id())
    Y = Y.withColumn('id', F.monotonically_increasing_id())
    joinedDF = X.join(Y,'id','inner')
    joinedDF = joinedDF.drop('id')
    return joinedDF
```

```python
#    7. Data splitting to training, testing, and validation
```

```python
def train_valid_test_split(df, train_size=0.4, valid_size=0.3,seed=12345):

    train, valid, test = df.randomSplit([train_size, valid_size,1-train_
    size-valid_size], seed=12345)
```

```
    return train,valid,test

#    8. Assembling vectors

def assembled_vectors(train,list_of_features_to_scale,target_column_name):

    stages = []
    assembler = VectorAssembler(inputCols=list_of_features_to_scale,
    outputCol='features')
    stages=[assembler]
    selectedCols = [target_column_name,'features'] + list_of_features_to_
    scale

    pipeline = Pipeline(stages=stages)
    assembleModel = pipeline.fit(train)

    train = assembleModel.transform(train).select(selectedCols)
    return train

#    9. Scaling input variables

def scaled_dataframes(train,valid,test,list_of_features_to_scale,target_
column_name):

    stages = []
    assembler = VectorAssembler(inputCols=list_of_features_to_scale,
    outputCol='assembled_features')
    scaler = StandardScaler(inputCol=assembler.getOutputCol(),
    outputCol='features')
    stages=[assembler,scaler]
    selectedCols = [target_column_name,'features'] + list_of_features_to_
    scale

    pipeline = Pipeline(stages=stages)
    pipelineModel = pipeline.fit(train)

    train = pipelineModel.transform(train).select(selectedCols)
    valid = pipelineModel.transform(valid).select(selectedCols)
```

```
    test = pipelineModel.transform(test).select(selectedCols)

    return train, valid, test, pipelineModel
```

Feature Selection

This module shortlists the variables and selects the top variables that are the most predictive features. It uses random forests to identify the top variables.

```
import pandas as pd
import matplotlib
matplotlib.use('Agg')
import matplotlib.pyplot as plt

# The module below is used to draw the feature importance plot
def draw_feature_importance(user_id, mdl_ltrl, importance_df):

    importance_df = importance_df.sort_values('Importance_Score')
    plt.figure(figsize=(15,15))
    plt.title('Feature Importances')
    plt.barh(range(len(importance_df['Importance_Score'])), importance_
    df['Importance_Score'], align='center')
    plt.yticks(range(len(importance_df['Importance_Score'])), importance_
    df['name'])
    plt.ylabel('Variable Importance')
    plt.savefig('/home/' + user_id + '/' + 'mla_' + mdl_ltrl + '/' +
    'Features selected for modeling.png', bbox_inches='tight')
    plt.close()
    return None

# The module below is used to save the feature importance as an Excel file
def save_feature_importance(user_id, mdl_ltrl, importance_df):
    importance_df.drop('idx',axis=1,inplace=True)
    importance_df = importance_df[0:30]
    importance_df.to_excel('/home/' + user_id + '/' + 'mla_' + mdl_ltrl +
    '/' + 'feature_importance.xlsx')
    draw_feature_importance(user_id, mdl_ltrl, importance_df)
```

```
    return None
```

The following module is used to calculate the feature importance for each
variable based on the Random Forest output. The feature importance is used
to reduce the final variable list to 30.

```
def ExtractFeatureImp(featureImp, dataset, featuresCol):
    """

    Takes in a feature importance from a random forest / GBT model and maps
    it to the column names
    Output as a pandas DataFrame for easy reading
    rf = RandomForestClassifier(featuresCol="features")
    mod = rf.fit(train)
    ExtractFeatureImp(mod.featureImportances, train, "features")
    """

    list_extract = []
    for i in dataset.schema[featuresCol].metadata["ml_attr"]["attrs"]:
        list_extract = list_extract + dataset.schema[featuresCol].
        metadata["ml_attr"]["attrs"][i]
    varlist = pd.DataFrame(list_extract)
    varlist['Importance_Score'] = varlist['idx'].apply(lambda x:
    featureImp[x])
    return(varlist.sort_values('Importance_Score', ascending = False))
```

Model Building

In this module, we define multiple algorithm functions that support binary targets,
including logistic regression, random forests, gradient boosting, and neural nets.

```
from pyspark.ml.classification import LogisticRegression
from pyspark.ml.classification import LogisticRegressionModel
# from sklearn.externals import joblib
import joblib

def logistic_model(train, x, y):
    lr = LogisticRegression(featuresCol = x, labelCol = y, maxIter = 10)
```

```
    lrModel = lr.fit(train)
    return lrModel

from pyspark.ml.classification import RandomForestClassifier
from pyspark.ml.classification import RandomForestClassificationModel

def randomForest_model(train, x, y):
    rf = RandomForestClassifier(featuresCol = x, labelCol = y, numTrees=10)
    rfModel = rf.fit(train)
    return rfModel

from pyspark.ml.classification import GBTClassifier
from pyspark.ml.classification import GBTClassificationModel

def gradientBoosting_model(train, x, y):
    gb = GBTClassifier(featuresCol = x, labelCol = y, maxIter=10)
    gbModel = gb.fit(train)
    return gbModel

from pyspark.ml.classification import DecisionTreeClassifier
from pyspark.ml.classification import DecisionTreeClassificationModel

def decisionTree_model(train, x, y):

    dt = DecisionTreeClassifier(featuresCol = x, labelCol = y, maxDepth=5)
    dtModel = dt.fit(train)
    return dtModel

from pyspark.ml.classification import MultilayerPerceptronClassifier
from pyspark.ml.classification import
MultilayerPerceptronClassificationModel

def neuralNetwork_model(train, x, y, feature_count):
    layers = [feature_count, feature_count*3, feature_count*2, 2]
    mlp = MultilayerPerceptronClassifier(featuresCol = x, labelCol = y,
    maxIter=100, layers=layers, blockSize=512,seed=12345)
    mlpModel = mlp.fit(train)
    return mlpModel
```

Metrics Calculation

The following module calculates the model metrics including KS, ROC, and accuracy.

```python
from pyspark.sql.types import DoubleType
from pyspark.sql import *
from pyspark.sql.functions import desc
from pyspark.sql.functions import udf
from pyspark.sql import functions as F
import sys
import time
# import __builtin__ as builtin
import builtins
from pyspark.ml.evaluation import BinaryClassificationEvaluator
from pyspark.ml.feature import QuantileDiscretizer
import numpy
import numpy as np
from pyspark import SparkContext,HiveContext,Row,SparkConf

spark = SparkSession.builder.appName("MLA_metrics_calculator").
enableHiveSupport().getOrCreate()
spark.sparkContext.setLogLevel('ERROR')
sc = spark.sparkContext

def highlight_max(data, color='yellow'):
    '''
    highlight the maximum in a Series or DataFrame
    '''
    attr = 'background-color: {}'.format(color)
    if data.ndim == 1:  # Series from .apply(axis=0) or axis=1
        is_max = data == data.max()
        return [attr if v else '' for v in is_max]
    else:  # from .apply(axis=None)
        is_max = data == data.max().max()
        return pd.DataFrame(np.where(is_max, attr, ''),index=data.index,
        columns=data.columns)

def calculate_metrics(predictions,y,data_type):
```

```
start_time4 = time.time()

# Calculate ROC
evaluator =
BinaryClassificationEvaluator(labelCol=y,rawPredictionCol='probability')
    auroc = evaluator.evaluate(predictions,{evaluator.metricName:
    "areaUnderROC"})
    print('AUC calculated',auroc)

    selectedCols = predictions.select(F.col("probability"),
F.col('prediction'), F.col(y)).rdd.map(lambda row:
(float(row['probability'][1]), float(row['prediction']), float(row[y]))).
collect()
    y_score, y_pred, y_true = zip(*selectedCols)

    # Calculate Accuracy
    accuracydf=predictions.withColumn('acc',F.when(predictions.
    prediction==predictions[y],1).otherwise(0))
    accuracydf.createOrReplaceTempView("accuracyTable")
    RFaccuracy=spark.sql("select sum(acc)/count(1) as accuracy from
    accuracyTable").collect()[0][0]
    print('Accuracy calculated',RFaccuracy)

#     # Build KS Table
    split1_udf = udf(lambda value: value[1].item(), DoubleType())

    if data_type in ['train','valid','test','oot1','oot2']:
        decileDF = predictions.select(y, split1_udf('probability').
        alias('probability'))
    else:
        decileDF = predictions.select(y, 'probability')

    decileDF=decileDF.withColumn('non_target',1-decileDF[y])

    window = Window.orderBy(desc("probability"))
    decileDF = decileDF.withColumn("rownum", F.row_number().over(window))
    decileDF.cache()
```

```
decileDF=decileDF.withColumn("rownum",decileDF["rownum"].
cast("double"))

window2 = Window.orderBy("rownum")
RFbucketedData=decileDF.withColumn("deciles", F.ntile(10).
over(window2))
RFbucketedData = RFbucketedData.withColumn('deciles',RFbucketedData['de
ciles'].cast("int"))
RFbucketedData.cache()
#a = RFbucketedData.count()
#print(RFbucketedData.show())

## to pandas from here
print('KS calculation starting')
target_cnt=RFbucketedData.groupBy('deciles').agg(F.sum(y).
alias('target')).toPandas()
non_target_cnt=RFbucketedData.groupBy('deciles').agg(F.sum("non_
target").alias('non_target')).toPandas()
overall_cnt=RFbucketedData.groupBy('deciles').count().alias('Total').
toPandas()
overall_cnt = overall_cnt.merge(target_cnt,on='deciles',how='inner').
merge(non_target_cnt,on='deciles',how='inner')
overall_cnt=overall_cnt.sort_values(by='deciles',ascending=True)
overall_cnt['Pct_target']=(overall_cnt['target']/overall_
cnt['count'])*100
overall_cnt['cum_target'] = overall_cnt.target.cumsum()
overall_cnt['cum_non_target'] = overall_cnt.non_target.cumsum()
overall_cnt['%Dist_Target'] = (overall_cnt['cum_target'] / overall_cnt.
target.sum())*100
overall_cnt['%Dist_non_Target'] = (overall_cnt['cum_non_target'] /
overall_cnt.non_target.sum())*100
overall_cnt['spread'] = builtins.abs(overall_cnt['%Dist_Target']-
overall_cnt['%Dist_non_Target'])
decile_table=overall_cnt.round(2)
print("KS_Value =", builtins.round(overall_cnt.spread.max(),2))
decileDF.unpersist()
```

```
RFbucketedData.unpersist()
print("Metrics calculation process Completed in : "+ " %s seconds" %
(time.time() - start_time4))
return auroc,RFaccuracy,builtins.round(overall_cnt.spread.max(),2),
y_score, y_pred, y_true, overall_cnt
```

Validation and Plot Generation

This module generates the various plots, including ROC, confusion matrix, and
KS. Model validation is also done as part of this module.

```
import matplotlib
matplotlib.use('Agg')
import matplotlib.pyplot as plt
from sklearn import metrics
import glob
import os
import pandas as pd
import seaborn as sns
from pandas import ExcelWriter
from metrics_calculator import *

# Generate ROC chart

def draw_roc_plot(user_id, mdl_ltrl, y_score, y_true, model_type, data_
type):

    fpr, tpr, thresholds = metrics.roc_curve(y_true, y_score, pos_label = 1)
    roc_auc = metrics.auc(fpr,tpr)
    plt.title(str(model_type) + ' Model - ROC for ' + str(data_type) + '
    data' )
    plt.plot([0, 1], [0, 1], 'r--')
    plt.plot(fpr, tpr, label = 'AUC = %0.2f' % roc_auc)
    plt.xlabel('False Positive Rate (1 - Specificity)')
    plt.ylabel('True Positive Rate (Sensitivity)')
```

```python
    plt.legend(loc = 'lower right')
    print('/home/' + user_id + '/' + 'mla_' + mdl_ltrl + '/' + str(model_
    type) + '/' + str(model_type) + ' Model - ROC for ' + str(data_type) +
    ' data.png')
    plt.savefig('/home/' + user_id + '/' + 'mla_' + mdl_ltrl + '/' +
    str(model_type) + '/' + str(model_type) + ' Model - ROC for ' +
    str(data_type) + ' data.png', bbox_inches='tight')
    plt.close()

# Generate KS chart


def draw_ks_plot(user_id, mdl_ltrl, model_type):

    writer = ExcelWriter('/home/' + user_id + '/' + 'mla_' + mdl_ltrl + '/'
    + str(model_type) + '/KS_Charts.xlsx')

    for filename in glob.glob('/home/' + user_id + '/' + 'mla_' + mdl_ltrl
    + '/' + str(model_type) + '/KS ' + str(model_type) + ' Model*.xlsx'):
        excel_file = pd.ExcelFile(filename)
        (_, f_name) = os.path.split(filename)
        (f_short_name, _) = os.path.splitext(f_name)
        for sheet_name in excel_file.sheet_names:
            df_excel = pd.read_excel(filename, sheet_name=sheet_name)
            df_excel = df_excel.style.apply(highlight_max,
            subset=['spread'], color='#e6b71e')
            df_excel.to_excel(writer, f_short_name, index=False)
            worksheet = writer.sheets[f_short_name]
            worksheet.conditional_format('C2:C11', {'type': 'data_
            bar','bar_color': '#34b5d9'})#,'bar_solid': True
            worksheet.conditional_format('E2:E11', {'type': 'data_
            bar','bar_color': '#366fff'})#,'bar_solid': True
        os.remove(filename)
    writer.save()

# Confusion matrix
```

```python
def draw_confusion_matrix(user_id, mdl_ltrl, y_pred, y_true, model_type,
data_type):

    AccuracyValue = metrics.accuracy_score(y_pred=y_pred, y_true=y_true)
    PrecisionValue = metrics.precision_score(y_pred=y_pred, y_true=y_true)
    RecallValue = metrics.recall_score(y_pred=y_pred, y_true=y_true)
    F1Value = metrics.f1_score(y_pred=y_pred, y_true=y_true)

    plt.title(str(model_type) + ' Model - Confusion
    Matrix for ' + str(data_type) + ' data \n \n
    Accuracy:{0:.3f}    Precision:{1:.3f}    Recall:{2:.3f}    F1
    Score:{3:.3f}\n'.format(AccuracyValue, PrecisionValue, RecallValue,
    F1Value))
    cm = metrics.confusion_matrix(y_true=y_true,y_pred=y_pred)
    sns.heatmap(cm, annot=True, fmt='g'); #annot=True to annotate cells
    plt.xlabel("Predicted labels")
    plt.ylabel("True labels")
    print('/home/' + user_id + '/' + 'mla_' + mdl_ltrl + '/' + str(model_
    type) + '/' + str(model_type) + ' Model - Confusion Matrix for ' +
    str(data_type) + ' data.png')
    plt.savefig('/home/' + user_id + '/' + 'mla_' + mdl_ltrl + '/' +
    str(model_type) + '/' + str(model_type) + ' Model - Confusion Matrix
    for ' + str(data_type) + ' data.png', bbox_inches='tight')
    plt.close()

# Model validation

def model_validation(user_id, mdl_ltrl, data, y, model, model_type, data_
type):

    start_time = time.time()

    pred_data = model.transform(data)
    print('model output predicted')

    roc_data, accuracy_data, ks_data, y_score, y_pred, y_true, decile_table
    = calculate_metrics(pred_data,y,data_type)
```

```
draw_roc_plot(user_id, mdl_ltrl, y_score, y_true, model_type, data_
type)
decile_table.to_excel('/home/' + user_id + '/' + 'mla_' + mdl_ltrl
+ '/' + str(model_type) + '/KS ' + str(model_type) + ' Model ' +
str(data_type) + '.xlsx',index=False)
draw_confusion_matrix(user_id, mdl_ltrl, y_pred, y_true, model_type,
data_type)
print('Metrics computed')

l = [roc_data, accuracy_data, ks_data]
end_time = time.time()
print("Model validation process completed in :  %s seconds" % (end_
time-start_time))
return l
```

Model Selection

This module is responsible for generating the challenger and champion model definitions based on validation metrics criteria. It also generates the consolidated Excel output file that contains the information of all models.

```
import pandas as pd
import joblib
import numpy as np
import glob
import os

def select_model(user_id, mdl_ltrl, model_selection_criteria, dataset_to_
use):
    df = pd.DataFrame({},columns=['roc_train', 'accuracy_train', 'ks_
    train', 'roc_valid', 'accuracy_valid', 'ks_valid', 'roc_test',
    'accuracy_test', 'ks_test', 'roc_oot1', 'accuracy_oot1', 'ks_oot1',
    'roc_oot2', 'accuracy_oot2', 'ks_oot2'])
    current_dir = os.getcwd()
    os.chdir('/home/' + user_id + '/' + 'mla_' + mdl_ltrl)
    for file in glob.glob('*metrics.z'):
        l = joblib.load(file)
```

```
        df.loc[str(file.split('_')[0])] = 1

for file in glob.glob('*metrics.z'):
    os.remove(file)

os.chdir(current_dir)
df.index = df.index.set_names(['model_type'])
df = df.reset_index()
model_selection_criteria = model_selection_criteria.lower()
column_to_sort = model_selection_criteria + '_' + dataset_to_use.
lower()
checker_value = 0.03

if model_selection_criteria == 'ks':
    checker_value = checker_value * 100

df['counter'] = (np.abs(df[column_to_sort] - df[model_
selection_criteria + '_train']) > checker_value).astype(int)
+                    (np.abs(df[column_to_sort] - df[model_
selection_criteria + '_valid']) > checker_value).astype(int)
+                    (np.abs(df[column_to_sort] - df[model_
selection_criteria + '_test']) > checker_value).astype(int)
+                    (np.abs(df[column_to_sort] - df[model_
selection_criteria + '_oot1']) > checker_value).astype(int)
+                    (np.abs(df[column_to_sort] - df[model_selection_
criteria + '_oot2']) > checker_value).astype(int)

df = df.sort_values(['counter', column_to_sort], ascending=[True,
False]).reset_index(drop=True)

df['selected_model'] = ''
df.loc[0,'selected_model'] = 'Champion'
df.loc[1,'selected_model'] = 'Challenger'

df.to_excel('/home/' + user_id + '/' + 'mla_' + mdl_ltrl + '/metrics.
xlsx')
return df
```

Score Code Creation

This module generates the pseudo score code for production deployment. It links all the model objects generated in the modeling process and places them in a single location. This also includes data manipulation and categorical variable processing functions. This file can be run independently as long as the model objects are saved in same location pointing to the right input data source.

```
#Import the scoring features
import string

import_packages = """
#This is a pseudo score code for production deployment. It links to all
your model objects created during the modeling process. If you plan to use
this file, then change the "score_table" variable to point to your input
data. Double-check the "home_path" and "hdfs_path" if you altered the
location of model objects.
import os
os.chdir('/home/jovyan/work/spark-warehouse/auto_model_builder')
from pyspark import SparkContext,HiveContext,Row,SparkConf
from pyspark.sql import *
from pyspark.ml import Pipeline
from pyspark.ml.linalg import Vectors,VectorUDT
from pyspark.sql.functions import *
from pyspark.mllib.stat import *
from pyspark.ml.feature import *
from pyspark.ml.feature import IndexToString,StringIndexer,VectorIndexer
from sklearn.metrics import roc_curve,auc
import numpy as np
import pandas as pd
import subprocess
from pyspark.ml.tuning import ParamGridBuilder, CrossValidator
from pyspark.ml import Pipeline,PipelineModel
from pyspark.sql import functions as func
from datetime import *
from pyspark.sql import SparkSession,SQLContext
from pyspark.sql.types import *
```

```
from dateutil.relativedelta import relativedelta
from data_manipulations import *
from model_builder import *
import datetime
from datetime import date
import string
import os
import sys
import time
import numpy
spark = SparkSession.builder.appName("MLA_Automated_Scorecode").
enableHiveSupport().getOrCreate()
spark.sparkContext.setLogLevel('ERROR')
sc = spark.sparkContext
"""

parameters = string.Template("""
user_id = '${user_id}'
mdl_output_id = '${mdl_output_id}'
mdl_ltrl = '${mdl_ltrl}'
#Since the hdfs and home path below are pointing to your user_id by
default, to use this file for scoring, you need to upload the model objects
in hdfs_path and home_path to the appropriate score location path (Could
be advanl or any other folder path). You would need the following files to
perform scoring.
#hdfs_path  - all the files in the path specified below
#home_path - 'model_scoring_info.z'
hdfs_path = '/user/${user_id}' + '/' + 'mla_${mdl_ltrl}' #update score
location hdfs_path
home_path = '/home/${user_id}' + '/' + 'mla_${mdl_ltrl}' #update score
location home_path
""")

import_variables = """
from sklearn.externals import joblib
from pyspark.ml import Pipeline,PipelineModel
```

```
final_vars,id_vars,vars_selected,char_vars,num_vars,impute_with,selected_
model,dev_table_name = joblib.load(home_path + '/model_scoring_info.z')
char_labels = PipelineModel.load(hdfs_path + '/char_label_model.h5')
pipelineModel = PipelineModel.load(hdfs_path + '/pipelineModel.h5')
"""

load_models = """
KerasModel = ''
loader_model_list = [LogisticRegressionModel,
RandomForestClassificationModel, GBTClassificationModel,
DecisionTreeClassificationModel, MultilayerPerceptronClassificationModel,
KerasModel]
models_to_run = ['logistic', 'randomForest','gradientBoosting','decisionTre
e','neuralNetwork','keras']
load_model = loader_model_list[models_to_run.index(selected_model)]
model = load_model.load(hdfs_path + '/' + selected_model + '_model.h5')
"""

score_function = """
score_table = spark.sql("select " + ", ".join(final_vars) + " from " + dev_
table_name) #update this query appropriately
def score_new_df(scoredf, model):
    newX = scoredf.select(final_vars)
    newX = newX.select(list(vars_selected))
    newX = char_labels.transform(newX)
    newX = numerical_imputation(newX,num_vars, impute_with)
    newX = newX.select([c for c in newX.columns if c not in char_vars])
    newX = rename_columns(newX, char_vars)
    finalscoreDF = pipelineModel.transform(newX)
    finalscoreDF.cache()
    finalpredictedDF = model.transform(finalscoreDF)
    finalpredictedDF.cache()
    return finalpredictedDF
ScoredDF = score_new_df(score_table, model)
"""
```

```python
def selected_model_scorecode(user_id, mdl_output_id, mdl_ltrl, parameters):

    parameters = parameters.substitute(locals())
    scorefile = open('/home/' + user_id + '/' + 'mla_' + mdl_ltrl + '/
    score_code_selected_model.py', 'w')
    scorefile.write(import_packages)
    scorefile.write(parameters)
    scorefile.write(import_variables)
    scorefile.write(load_models)
    scorefile.write(score_function)
    scorefile.close()
    print('Score code generation complete')

# # Generate individual score codes

def individual_model_scorecode(user_id, mdl_output_id, mdl_ltrl,
parameters):

    loader_model_list = ['LogisticRegressionModel',
'RandomForestClassificationModel',
'GBTClassificationModel', 'DecisionTreeClassificationModel',
'MultilayerPerceptronClassificationModel', 'KerasModel']
    models_to_run = ['logistic', 'randomForest','gradientBoosting','decisio
    nTree','neuralNetwork','keras']

    parameters = parameters.substitute(locals())
    for i in models_to_run:
        try:
            load_model = loader_model_list[models_to_run.index(i)]

            write_model_parameter = string.Template("""
model = ${load_model}.load(hdfs_path + '/' + ${i} + '_model.h5')
            """).substitute(locals())

            scorefile = open('/home/' + user_id + '/' + 'mla_' + mdl_ltrl
            + '/' + str(i[0].upper()) + str(i[1:]) + '/score_code_' + i +
            '_model.py', 'w')
```

```
            scorefile.write(import_packages)
            scorefile.write(parameters)
            scorefile.write(import_variables)
            scorefile.write(write_model_parameter)
            scorefile.write(score_function)
            scorefile.close()
        except:
            pass

    print('Individual Score code generation complete')
```

Collating Results

A zipfile module is used to here to bind all the output generated by the earlier modules into a single zip file. This file can be easily exported for deploying models elsewhere.

```
import os
import zipfile

def retrieve_file_paths(dirName):

  # set up filepaths variable
    filePaths = []

  # Read all directory, subdirectories, and file lists
    for root, directories, files in os.walk(dirName):
        for filename in files:
        # Create the full filepath by using os module.
            filePath = os.path.join(root, filename)
            filePaths.append(filePath)

    # return all paths
    return filePaths

# Declare the main function
def zipper(dir_name):
# Assign the name of the directory to zip
```

```
# Call the function to retrieve all files and folders of the assigned
directory
filePaths = retrieve_file_paths(dir_name)

# printing the list of all files to be zipped
print('The following list of files will be zipped:')
for fileName in filePaths:
    print(fileName)

# writing files to a zipfile
zip_file = zipfile.ZipFile(dir_name+'.zip', 'w')
with zip_file:
    for file in filePaths:
        zip_file.write(file)

print(dir_name+'.zip file is created successfully!')
return(dir_name+'.zip')
```

Framework

This framework file combines all eight preceding modules to orchestrate them in a logical flow to create the desired output. It treats the data and performs variable selection. It builds machine learning algorithms and validates the models on holdout datasets. It picks the best algorithm based on user-selected statistics. It also produces the scoring code for production.

This framework file receives all the inputs. We save all the preceding modules, including this framework file, as a Python file. Since we are dealing with a churn modeling dataset here, we provide the input csv file location in the data_folder_path. All the other required inputs are provided in the first block of the code. All this code is designed to accommodate the Docker version of PySpark. By making minimal changes, this can be adapted to any cluster execution as well.

```
from pyspark import SparkContext,HiveContext,Row,SparkConf
from pyspark.sql import *
from pyspark.ml import Pipeline
from pyspark.ml.linalg import Vectors,VectorUDT
from pyspark.sql.functions import *
```

```
from pyspark.mllib.stat import *
from pyspark.ml.feature import *
from pyspark.ml.feature import IndexToString,StringIndexer,VectorIndexer
from sklearn.metrics import roc_curve,auc
import numpy as np
import pandas as pd
import subprocess
from pyspark.ml.tuning import ParamGridBuilder, CrossValidator
from pyspark.ml import Pipeline,PipelineModel
from pyspark.sql import functions as func
from datetime import *
from pyspark.sql import SparkSession,SQLContext
from pyspark.sql.types import *
from dateutil.relativedelta import relativedelta
import datetime
from datetime import date
import string
import os
import sys
import time
import numpy

spark = SparkSession.builder.appName("Automated_model_building").
enableHiveSupport().getOrCreate()
spark.sparkContext.setLogLevel('ERROR')
sc = spark.sparkContext

import_data = False
stop_run = False
message = ''
filename = ''

user_id = 'jovyan'
mdl_output_id = 'test_run01' #An unique ID to represent the model
mdl_ltrl = 'chapter8_testrun' #An unique literal or tag to represent the
model
```

```
input_dev_file='churn_modeling.csv'
input_oot1_file=''
input_oot2_file=''

dev_table_name = ''
oot1_table_name = ''
oot2_table_name = ''

delimiter_type = ','

include_vars = '' # user specified variables to be used
include_prefix = '' # user specified prefixes to be included for modeling
include_suffix = '' # user specified prefixes to be included for modeling
exclude_vars = 'rownumber,customerid,surname' # user specified variables to
be excluded for modeling
exclude_prefix = '' # user specified prefixes to be excluded for modeling
exclude_suffix = '' # user specified suffixes to be excluded for modeling

target_column_name = 'exited'

run_logistic_model = 1
run_randomforest_model = 1
run_boosting_model = 1
run_neural_model = 1

miss_per = 0.75
impute_with = 0.0
train_size=0.7
valid_size=0.2
seed=2308

model_selection_criteria = 'ks' #possible_values ['ks','roc','accuracy']
dataset_to_use = 'train' #possible_values ['train','valid','test','oot1','o
ot2']

data_folder_path = '/home/jovyan/work/'
hdfs_folder_path = '/home/jovyan/work/spark-warehouse/'

######################################################################
```

```
######No input changes required below this for default run##########
####################################################################

if input_oot1_file=='':
    input_oot1_file=input_dev_file
if input_oot2_file=='':
    input_oot2_file=input_dev_file
# assign input files if the user uploaded files instead of tables.
if dev_table_name.strip() == '':
    dev_input_file = input_dev_file
    if dev_input_file.strip() == '':
        print('Please provide a development table or development file to
        process the application')
        stop_run = True
        message = 'Development Table or file is not provided. Please
        provide a development table or file name to process'

    import_data = True
    file_type = dev_table_name.split('.')[-1]
    out,err=subprocess.Popen(['cp',data_folder_path+dev_input_file,hdfs_
    folder_path],stdout=subprocess.PIPE,stderr=subprocess.PIPE).communicate()

if oot1_table_name.strip() == '':
    oot1_input_file = input_oot1_file
    out,err=subprocess.Popen(['cp',data_folder_path+oot1_input_file,hdfs_
    folder_path],stdout=subprocess.PIPE,stderr=subprocess.PIPE).communicate()

if oot2_table_name.strip() == '':
    oot2_input_file = input_oot2_file
    out,err=subprocess.Popen(['cp',data_folder_path+oot2_input_file,hdfs_
    folder_path],stdout=subprocess.PIPE,stderr=subprocess.PIPE).communicate()

ignore_data_type = ['timestamp', 'date']
ignore_vars_based_on_datatype = []

# extract the input variables in the file or table
if not stop_run:
    if import_data:
```

```
        df = spark.read.option("delimiter",delimiter_type).option("header",
        "true").option("inferSchema", "true").csv(hdfs_folder_path + dev_
        input_file)
        df = pd.DataFrame(zip(*df.dtypes),['col_name', 'data_type']).T
else:
        df = spark.sql('describe ' + dev_table_name)
        df = df.toPandas()

input_vars = list(str(x.lower()) for x in df['col_name'])
print(input_vars)
for i in ignore_data_type:
        ignore_vars_based_on_datatype += list(str(x) for x in df[df['data_
        type'] == i]['col_name'])

if len(ignore_vars_based_on_datatype) > 0:
        input_vars = list(set(input_vars) - set(ignore_vars_based_on_
        datatype))

input_vars.remove(target_column_name)


## variables to include
import re
prefix_include_vars = []
suffix_include_vars = []

if include_vars.strip() != '':
        include_vars = re.findall(r'\w+', include_vars.lower())

if include_prefix.strip() != '':
        prefix_to_include = re.findall(r'\w+', include_prefix.lower())

        for i in prefix_to_include:
                temp = [x for x in input_vars if x.startswith(str(i))]
                prefix_include_vars.append(temp)

        prefix_include_vars = [item for sublist in prefix_include_vars for
        item in sublist]
```

```python
    if include_suffix.strip() != '':
        suffix_to_include = re.findall(r'\w+', include_suffix.lower())

        for i in suffix_to_include:
            temp = [x for x in input_vars if x.startswith(str(i))]
            suffix_include_vars.append(temp)

        suffix_include_vars = [item for sublist in suffix_include_vars for
        item in sublist]

    include_list = list(set(include_vars) | set(prefix_include_vars) |
    set(suffix_include_vars))


    ## Variables to exclude
    prefix_exclude_vars = []
    suffix_exclude_vars = []

    if exclude_vars.strip() != '':
        exclude_vars = re.findall(r'\w+', exclude_vars.lower())

    if exclude_prefix.strip() != '':
        prefix_to_exclude = re.findall(r'\w+', exclude_prefix.lower())

        for i in prefix_to_exclude:
            temp = [x for x in input_vars if x.startswith(str(i))]
            prefix_exclude_vars.append(temp)

        prefix_exclude_vars = [item for sublist in prefix_exclude_vars for
        item in sublist]

    if exclude_suffix.strip() != '':
        suffix_to_exclude = re.findall(r'\w+', exclude_suffix.lower())

        for i in suffix_to_exclude:
            temp = [x for x in input_vars if x.startswith(str(i))]
            suffix_exclude_vars.append(temp)

        suffix_exclude_vars = [item for sublist in suffix_exclude_vars for
        item in sublist]
```

```python
    exclude_list = list(set(exclude_vars) | set(prefix_exclude_vars) |
    set(suffix_exclude_vars))

    if len(include_list) > 0:
        input_vars = list(set(input_vars) & set(include_list))

    if len(exclude_list) > 0:
        input_vars = list(set(input_vars) - set(exclude_list))

if not stop_run:

    final_vars = input_vars  # final list of variables to be pulled
    from datetime import datetime
    insertion_date = datetime.now().strftime("%Y-%m-%d")

    import re
    from pyspark.sql.functions import col

    # import data for the modeling
    if import_data:
        train_table = spark.read.option("delimiter",delimiter_type).
        option("header", "true").option("inferSchema", "true").csv(hdfs_
        folder_path + dev_input_file)
        oot1_table = spark.read.option("delimiter",delimiter_type).
        option("header", "true").option("inferSchema", "true").csv(hdfs_
        folder_path + oot1_input_file)
        oot2_table = spark.read.option("delimiter",delimiter_type).
        option("header", "true").option("inferSchema", "true").csv(hdfs_
        folder_path + oot2_input_file)
    else:
        train_table = spark.sql("select " + ", ".join(final_vars + [target_
        column_name]) + " from " + dev_table_name)
        oot1_table = spark.sql("select " + ", ".join(final_vars + [target_
        column_name]) + " from " + oot1_table_name)
        oot2_table = spark.sql("select " + ", ".join(final_vars + [target_
        column_name]) + " from " + oot2_table_name)
```

345

```
train_table = train_table.where(train_table[target_column_name].
isNotNull())
oot1_table = oot1_table.where(oot1_table[target_column_name].
isNotNull())
oot2_table = oot2_table.where(oot2_table[target_column_name].
isNotNull())
print (final_vars)

oot1_table=oot1_table.toDF(*[c.lower() for c in oot1_table.columns])
oot2_table=oot2_table.toDF(*[c.lower() for c in oot2_table.columns])
print(oot1_table.columns)
print(oot2_table.columns)
X_train = train_table.select(*final_vars)
X_train.cache()

# apply data manipulations on the data - missing value check, label
encoding, imputation

from data_manipulations import *

vars_selected_train = missing_value_calculation(X_train, miss_per) #
missing value check

vars_selected = filter(None,list(set(list(vars_selected_train))))
print('vars selected')
X = X_train.select(*vars_selected)
print(X.columns)
vars_selectedn=X.columns
X = X.cache()

Y = train_table.select(target_column_name)
Y = Y.cache()

char_vars, num_vars = identify_variable_type(X)
X, char_labels = categorical_to_index(X, char_vars) #label encoding
X = numerical_imputation(X,num_vars, impute_with) # imputation
X = X.select([c for c in X.columns if c not in char_vars])
```

```
X = rename_columns(X, char_vars)
joinedDF = join_features_and_target(X, Y)

joinedDF = joinedDF.cache()
print('Features and targets are joined')

train, valid, test = train_valid_test_split(joinedDF, train_size,
valid_size, seed)
train = train.cache()
valid = valid.cache()
test = test.cache()
print('Train, valid and test dataset created')

x = train.columns
x.remove(target_column_name)
feature_count = len(x)
print(feature_count)

if feature_count > 30:
    print('# No of features - ' + str(feature_count) + '.,  Performing
    feature reduction before running the model.')

# directory to produce the outputs of the automation
import os

try:
    if not os.path.exists('/home/' + user_id + '/' + 'mla_' + mdl_
    ltrl):
        os.mkdir('/home/' + user_id + '/' + 'mla_' + mdl_ltrl)
except:
    user_id = 'jovyan'
    if not os.path.exists('/home/' + user_id + '/' + 'mla_' + mdl_
    ltrl):
        os.mkdir('/home/' + user_id + '/' + 'mla_' + mdl_ltrl)

subprocess.call(['chmod','777','-R','/home/' + user_id + '/' + 'mla_' +
mdl_ltrl])


x = train.columns
```

```
x.remove(target_column_name)
sel_train = assembled_vectors(train,x, target_column_name)
sel_train.cache()

# # Variable Reduction for more than 30 variables in the feature set
using Random Forest

from pyspark.ml.classification import  RandomForestClassifier
from feature_selection import *

rf = RandomForestClassifier(featuresCol="features",labelCol = target_
column_name)
mod = rf.fit(sel_train)
varlist = ExtractFeatureImp(mod.featureImportances, sel_train,
"features")
selected_vars = [str(x) for x in varlist['name'][0:30]]
train = train.select([target_column_name] + selected_vars)
train.cache()

save_feature_importance(user_id, mdl_ltrl, varlist) #Create feature
importance plot and excel data

x = train.columns
x.remove(target_column_name)
feature_count = len(x)
print(feature_count)

train, valid, test, pipelineModel = scaled_dataframes(train,valid,test,
x,target_column_name)

train = train.cache()
valid = valid.cache()
test = test.cache()
print('Train, valid and test are scaled')
print (train.columns)

# import packages to perform model building, validation, and plots
```

```
import time
from validation_and_plots import *

# apply the transformation done on training dataset to OOT 1 and OOT 2
using the score_new_df function
def score_new_df(scoredf):
    newX = scoredf.select(*final_vars)
    #idX = scoredf.select(id_vars)
    print(newX.columns)
    newX = newX.select(*vars_selectedn)
    print(newX.columns)
    newX = char_labels.transform(newX)
    newX = numerical_imputation(newX,num_vars, impute_with)
    newX = newX.select([c for c in newX.columns if c not in char_vars])
    newX = rename_columns(newX, char_vars)

    finalscoreDF = pipelineModel.transform(newX)
    finalscoreDF.cache()
    return finalscoreDF

# apply the transformation done on training dataset to OOT 1 and OOT 2
using the score_new_df function

x = 'features'
y = target_column_name

oot1_targetY = oot1_table.select(target_column_name)
print(oot1_table.columns)
oot1_intDF = score_new_df(oot1_table)
oot1_finalDF = join_features_and_target(oot1_intDF, oot1_targetY)
oot1_finalDF.cache()
print(oot1_finalDF.dtypes)

oot2_targetY = oot2_table.select(target_column_name)
oot2_intDF = score_new_df(oot2_table)
oot2_finalDF = join_features_and_target(oot2_intDF, oot2_targetY)
oot2_finalDF.cache()
print(oot2_finalDF.dtypes)
```

```python
# run individual models

from model_builder import *
from metrics_calculator import *

loader_model_list = []
dataset_list = ['train','valid','test','oot1','oot2']
datasets = [train,valid,test,oot1_finalDF, oot2_finalDF]
print(train.count())
print(test.count())
print(valid.count())
print(oot1_finalDF.count())
print(oot2_finalDF.count())
models_to_run = []

if run_logistic_model:
    lrModel = logistic_model(train, x, y) #build model
    lrModel.write().overwrite().save('/home/' + user_id + '/' + 'mla_'
    + mdl_ltrl + '/logistic_model.h5') #save model object
    print("Logistic model developed")
    model_type = 'Logistic'
    l = []

    try:
        os.mkdir('/home/' + user_id + '/' + 'mla_' + mdl_ltrl + '/' +
        str(model_type))
    except:
        pass

    for i in datasets:
        l += model_validation(user_id, mdl_ltrl, i, y, lrModel, model_
        type, dataset_list[datasets.index(i)]) #validate model

    draw_ks_plot(user_id, mdl_ltrl, model_type) #ks charts
    joblib.dump(l,'/home/' + user_id + '/' + 'mla_' + mdl_ltrl  + '/
    logistic_metrics.z') #save model metrics
    models_to_run.append('logistic')
    loader_model_list.append(LogisticRegressionModel)
```

```python
if run_randomforest_model:
    rfModel = randomForest_model(train, x, y) #build model
    rfModel.write().overwrite().save('/home/' + user_id + '/' + 'mla_'
    + mdl_ltrl + '/randomForest_model.h5') #save model object
    print("Random Forest model developed")
    model_type = 'RandomForest'
    l = []

    try:
        os.mkdir('/home/' + user_id + '/' + 'mla_' + mdl_ltrl + '/' +
        str(model_type))
    except:
        pass

    for i in datasets:
        l += model_validation(user_id, mdl_ltrl, i, y, rfModel, model_
        type, dataset_list[datasets.index(i)]) #validate model

    draw_ks_plot(user_id, mdl_ltrl, model_type) #ks charts
    joblib.dump(l,'/home/' + user_id + '/' + 'mla_' + mdl_ltrl + '/
    randomForest_metrics.z') #save model metrics
    models_to_run.append('randomForest')
    loader_model_list.append(RandomForestClassificationModel)

if run_boosting_model:
    gbModel = gradientBoosting_model(train, x, y) #build model
    gbModel.write().overwrite().save('/home/' + user_id + '/' + 'mla_'
    + mdl_ltrl + '/gradientBoosting_model.h5') #save model object
    print("Gradient Boosting model developed")
    model_type = 'GradientBoosting'
    l = []

    try:
        os.mkdir('/home/' + user_id + '/' + 'mla_' + mdl_ltrl + '/' +
        str(model_type))
    except:
        pass
```

```
    for i in datasets:
        l += model_validation(user_id, mdl_ltrl, i, y, gbModel, model_
        type, dataset_list[datasets.index(i)]) #validate model

    draw_ks_plot(user_id, mdl_ltrl, model_type) #ks charts
    joblib.dump(l,'/home/' + user_id + '/' + 'mla_' + mdl_ltrl + '/
    gradientBoosting_metrics.z') #save model metrics
    models_to_run.append('gradientBoosting')
    loader_model_list.append(GBTClassificationModel)

if run_neural_model:
    mlpModel = neuralNetwork_model(train, x, y, feature_count) #build
    model
    mlpModel.write().overwrite().save('/home/' + user_id + '/' + 'mla_'
    + mdl_ltrl + '/neuralNetwork_model.h5') #save model object
    print("Neural Network model developed")
    model_type = 'NeuralNetwork'
    l = []

    try:
        os.mkdir('/home/' + user_id + '/' + 'mla_' + mdl_ltrl + '/' +
        str(model_type))
    except:
        pass

    for i in datasets:
        l += model_validation(user_id, mdl_ltrl, i, y, mlpModel, model_
        type, dataset_list[datasets.index(i)]) #validate model

    draw_ks_plot(user_id, mdl_ltrl, model_type) #ks charts
    joblib.dump(l,'/home/' + user_id + '/' + 'mla_' + mdl_ltrl + '/
    neuralNetwork_metrics.z') #save model metrics
    models_to_run.append('neuralNetwork')
    loader_model_list.append(MultilayerPerceptronClassificationModel)

# model building complete. Let us validate the metrics for the models
created
```

```
# model validation part starts now.
from model_selection import *
output_results = select_model(user_id, mdl_ltrl, model_selection_
criteria, dataset_to_use) #select Champion, Challenger based on the
metrics provided by user

#print(type(output_results), output_results)

selected_model = output_results['model_type'][0] #Champion model based
on selected metric

load_model = loader_model_list[models_to_run.index(selected_model)]
#load the model object for Champion model
model = load_model.load('/home/' + user_id + '/' + 'mla_' + mdl_ltrl +
'/' + selected_model + '_model.h5')

print('Model selected for scoring - ' + selected_model)

# Produce pseudo score for production deployment
# save objects produced in the steps above for future scoring
import joblib

char_labels.write().overwrite().save('/home/' + user_id + '/' + 'mla_'
+ mdl_ltrl + '/char_label_model.h5')
pipelineModel.write().overwrite().save('/home/' + user_id + '/' +
'mla_' + mdl_ltrl + '/pipelineModel.h5')

save_list = [final_vars,vars_selected,char_vars,num_vars,impute_
with,selected_model,dev_table_name]
joblib.dump(save_list,'/home/' + user_id + '/' + 'mla_' + mdl_ltrl + '/
model_scoring_info.z')

# # Create score code

from scorecode_creator import *
selected_model_scorecode(user_id, mdl_output_id, mdl_ltrl, parameters)
individual_model_scorecode(user_id, mdl_output_id, mdl_ltrl,
parameters)
```

```
    message = message + 'Model building activity complete and the results
    are attached with this email. Have Fun'

    from zipper_function import *
    try:
        filename = zipper('/home/' + user_id + '/' + 'mla_' + mdl_ltrl)
    except:
        filename = ''
# clean up files loaded in the local path
if import_data:
    file_list = [dev_input_file, oot1_input_file, oot2_input_file]

    for i in list(set(file_list)):
        try:
            os.remove(data_folder_path + str(i))
        except:
            pass

# clean up files loaded in the hdfs path
if import_data:
    file_list = [dev_input_file, oot1_input_file, oot2_input_file]

    for i in list(set(file_list)):
        try:
            out,err=subprocess.Popen([ 'rm','-r','-f',hdfs_folder_path+str(i)],
                stdout=subprocess.PIPE,stderr=subprocess.PIPE).communicate()
        except:
            pass
```

For execution, we will open Docker using the following command:

```
docker run -it -p 8888:8888 -v /Users/ramcharankakarla/demo_data/:/home/
jovyan/work/ jupyter/pyspark-notebook:latest bash
```

On the local machine, copy all the files to the path exposed to Docker (e.g., /Users/
ramcharankakarla/demo_data/). We have created a folder *Chapter8_automator* and
placed all the required files there (Table 8-3). Also copy the *churn_modeling.csv* dataset
to the same folder.

Table 8-3. *Files and their functions in the*
Automation Framework

Module	Filename
Data manipulations	data_manipulations.py
Feature selection	feature_selection.py
Model building	model_builder.py
Metrics calculation	metrics_calculator.py
Validation and plot generation	validation_and_plots.py
Model selection	model_selection.py
Score code creation	scorecode_creator.py
Collating results	zipper_function.py
Framework	build_and_execute_pipe.py

Before running these framework files, make sure to install or check for the existence of the following packages in Docker:

```
pip install openpyxl
```

```
pip install xlsxwriter
```

To run the framework file from Docker, use the following command:

```
spark-submit --master local[*] /home/jovyan/work/Chapter8_automator/build_
and_execute_pipe.py
```

After the execution is completed, all the data will be stored in /home/jovyan/.

To make a local copy of the files and metrics and prevent loss of any work, use the following command. This creates a local copy of the work. Similarly, you can copy a zip file (*mla_chapter8_testrun.zip*) if needed.

```
cp -r mla_chapter8_testrun /home/jovyan/work/
```

Nice job! We have successfully built an end-to-end automation engine that can save you a significant amount of time. Let's take a look at the output this engine has generated.

Pipeline Outputs

Figures 8-13 through 8-15 and Tables 8-4 through 8-6 illustrate the outputs generated by the automated pipeline framework. The following outputs are generated for any given binary model.

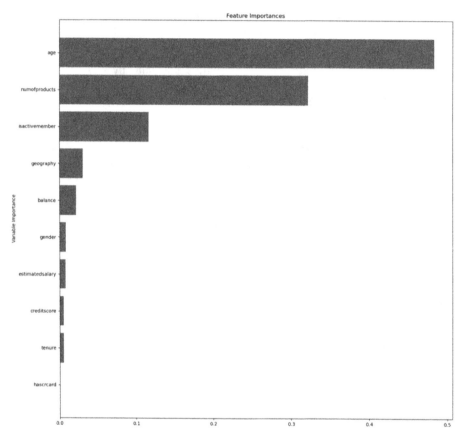

Figure 8-13. *Automated model feature importances*

Table 8-4. *Combined Metrics*

model_type	roc_train	accuracy_train	ks_train	roc_valid	accuracy_valid	ks_valid
neuralNetwork	0.86171092	0.858369099	55.5	0.857589	0.862000986	54.43
randomForest	0.838162018	0.860228898	50.57	0.835679	0.856086742	50.48
logistic	0.755744261	0.81230329	37.19	0.754326	0.807294234	37.31
gradientBoosting	0.881999528	0.869384835	59.46	0.858617	0.853622474	55.95

model_type	roc_test	accuracy_test	ks_test	roc_oot1	accuracy_oot1
neuralNetwork	0.813792	0.849134	46.58	0.856175	0.8582
randomForest	0.806239	0.840979	44.3	0.834677	0.8575
logistic	0.719284	0.805301	32.19	0.751941	0.8106
gradientBoosting	0.816005	0.850153	47.22	0.870804	0.8643

model_type	ks_oot1	roc_oot2	accuracy_oot2	ks_oot2	selected_model
neuralNetwork	54.43	0.856175108	0.8582	54.43	Champion
randomForest	49.99	0.834676561	0.8575	49.99	Challenger
logistic	36.49	0.751941185	0.8106	36.49	
gradientBoosting	57.51	0.870803793	0.8643	57.51	

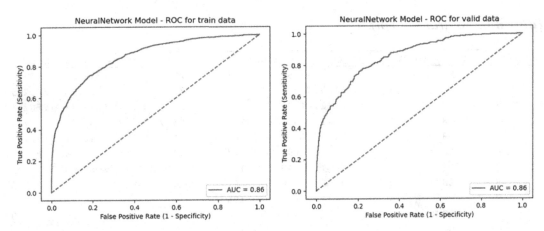

Figure 8-14. *Automated model ROCs*

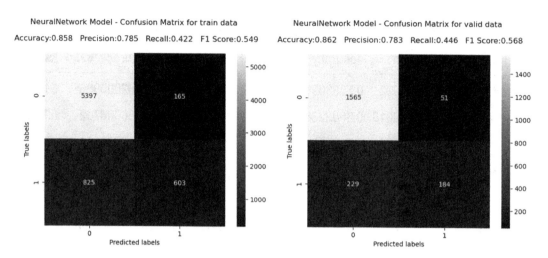

Figure 8-15. *Automated model — Confusion matrix*

Table 8-5. *Kolmogorov-Smirnov Statistic*

deciles	count	target	non_target	Pct_target	cum_target	cum_non_targ	Dist_Targe	ist_non_Tar	spread
1	699	577	122	82.54649	577	122	40.40616	2.193456	38.21271
2	699	299	400	42.77539	876	522	61.34454	9.385113	51.95942
3	699	183	516	26.18026	1059	1038	74.15966	18.66235	55.49731
4	699	113	586	16.16595	1172	1624	82.07283	29.19813	52.8747
5	699	99	600	14.16309	1271	2224	89.0056	39.98562	49.01999
6	699	67	632	9.585122	1338	2856	93.69748	51.34844	42.34904
7	699	32	667	4.577969	1370	3523	95.93838	63.34052	32.59785
8	699	33	666	4.72103	1403	4189	98.2493	75.31464	22.93466
9	699	12	687	1.716738	1415	4876	99.08964	87.66631	11.42333
10	699	13	686	1.8598	1428	5562	100	100	0

Table 8-6. *Neural Network Test*

deciles	count	target	non_target	Pct_target	cum_target	cum_non_targ	Dist_Targe	ist_non_Tar	spread
1	99	73	26	73.73737	73	26	37.2449	3.312102	33.9328
2	98	36	62	36.73469	109	88	55.61224	11.21019	44.40205
3	98	23	75	23.46939	132	163	67.34694	20.76433	46.58261
4	98	13	85	13.26531	145	248	73.97959	31.59236	42.38724
5	98	15	83	15.30612	160	331	81.63265	42.16561	39.46705
6	98	17	81	17.34694	177	412	90.30612	52.48408	37.82205
7	98	11	87	11.22449	188	499	95.91837	63.56688	32.35149
8	98	3	95	3.061224	191	594	97.44898	75.66879	21.78019
9	98	4	94	4.081633	195	688	99.4898	87.64331	11.84648
10	98	1	97	1.020408	196	785	100	100	0

This module is intended to help you create quick, automated experiments and is in no way meant to replace the model-building activity.

EXERCISE 8-2: BUILDING CUSTOM PIPELINES

Question: We challenge you to build a separate pipeline or integrate code flow to accommodate continuous targets.

Summary

- We learned the challenges of model management and deployment.

- We now know how to use MLflow for managing experiments and deploying models.

- We explored how to build custom pipelines for various model-building activities.

- We saw how pipelines can be stacked together to create an automated pipeline.

Great job! You are now familiar with some of the key concepts that will be useful in putting a model into production. This should give you a fair idea of how you want to manage a model lifecycle as you are building your data-preparation pipelines. In the next chapter, we will cover some of the tips, tricks, and interesting topics that can be useful in your day-to-day work.

CHAPTER 9

Deploying Machine Learning Models

This chapter will educate you on best practices to follow when deploying a machine learning model. We will use three methods to demonstrate the production deployment, as follows:

- Model deployment using HDFS object and pickle files

- Model deployment using Docker

- Real-time scoring API

In the previous chapter, you got a glimpse of the first technique. In this section, we will cover it in more detail. The code provided is the starter code to build the model and evaluate the results.

Starter Code

This code is just a template script for you to load the *bank-full.csv* file and build a model. If you already have a model, you can skip this section.

```
#default parameters
filename = "bank-full.csv"
target_variable_name = "y"

#load datasets
from pyspark.sql import SparkSession
spark = SparkSession.builder.getOrCreate()
df = spark.read.csv(filename, header=True, inferSchema=True, sep=';')
df.show()
```

© Ramcharan Kakarla, Sundar Krishnan and Sridhar Alla 2021
R. Kakarla et al., *Applied Data Science Using PySpark*, https://doi.org/10.1007/978-1-4842-6500-0_9

```
#convert target column variable to numeric
from pyspark.sql import functions as F
df = df.withColumn('y', F.when(F.col("y") == 'yes', 1).otherwise(0))

#datatypes of each column and treat them accordingly
# identify variable types function
def variable_type(df):
    vars_list = df.dtypes
    char_vars = []
    num_vars = []
    for i in vars_list:
        if i[1] in ('string'):
            char_vars.append(i[0])
        else:
            num_vars.append(i[0])
    return char_vars, num_vars

char_vars, num_vars = variable_type(df)
num_vars.remove(target_variable_name)

from pyspark.ml.feature import StringIndexer
from pyspark.ml import Pipeline

# convert categorical to numeric using label encoder option
def category_to_index(df, char_vars):

    char_df = df.select(char_vars)
    indexers = [StringIndexer(inputCol=c, outputCol=c+"_index",
    handleInvalid="keep") for c in char_df.columns]
    pipeline = Pipeline(stages=indexers)
    char_labels = pipeline.fit(char_df)
    df = char_labels.transform(df)
    return df, char_labels

df, char_labels = category_to_index(df, char_vars)
df = df.select([c for c in df.columns if c not in char_vars])

# rename encoded columns to original variable name
def rename_columns(df, char_vars):
```

```
    mapping = dict(zip([i + '_index' for i in char_vars], char_vars))
    df = df.select([F.col(c).alias(mapping.get(c, c)) for c in df.columns])
    return df

df = rename_columns(df, char_vars)

from pyspark.ml.feature import VectorAssembler

#assemble individual columns to one column - 'features'
def assemble_vectors(df, features_list, target_variable_name):
    stages = []
    #assemble vectors
    assembler = VectorAssembler(inputCols=features_list,
    outputCol='features')
    stages = [assembler]
    #select all the columns + target + newly created 'features' column
    selectedCols = [target_variable_name, 'features'] + features_list
    #use pipeline to process sequentially
    pipeline = Pipeline(stages=stages)
    #assembler model
    assembleModel = pipeline.fit(df)
    #apply assembler model on data
    df = assembleModel.transform(df).select(selectedCols)

    return df, assembleModel, selectedCols


#exclude target variable and select all other feature vectors
features_list = df.columns
#features_list = char_vars #this option is used only for ChiSqselector
features_list.remove(target_variable_name)

# apply the function on our DataFrame
df, assembleModel, selectedCols = assemble_vectors(df, features_list,
target_variable_name)

#train test split - 70/30 split
train, test = df.randomSplit([0.7, 0.3], seed=12345)
```

```
train.count(), test.count()

# Random forest model
from pyspark.ml.classification import RandomForestClassifier

clf = RandomForestClassifier(featuresCol='features', labelCol='y')
clf_model = clf.fit(train)
print(clf_model.featureImportances)
print(clf_model.toDebugString)
train_pred_result = clf_model.transform(train)
test_pred_result = clf_model.transform(test)

# Validate random forest model
from pyspark.mllib.evaluation import MulticlassMetrics
from pyspark.ml.evaluation import BinaryClassificationEvaluator
from pyspark.sql.types import IntegerType, DoubleType

def evaluation_metrics(df, target_variable_name):
    pred = df.select("prediction", target_variable_name)
    pred = pred.withColumn(target_variable_name, pred[target_variable_
            name].cast(DoubleType()))
    pred = pred.withColumn("prediction", pred["prediction"].
            cast(DoubleType()))
    metrics = MulticlassMetrics(pred.rdd.map(tuple))
    # confusion matrix
    cm = metrics.confusionMatrix().toArray()
    acc = metrics.accuracy #accuracy
    misclassification_rate = 1 - acc #misclassification rate
    precision = metrics.precision(1.0) #precision
    recall = metrics.recall(1.0) #recall
    f1 = metrics.fMeasure(1.0) #f1-score
    #roc value
    evaluator_roc = BinaryClassificationEvaluator
                    (labelCol=target_variable_name, rawPredictionCol='rawPr
                    ediction', metricName='areaUnderROC')
    roc = evaluator_roc.evaluate(df)
```

```
        evaluator_pr = BinaryClassificationEvaluator
                    (labelCol=target_variable_name, rawPredictionCol='rawPre
                    diction', metricName='areaUnderPR')
        pr = evaluator_pr.evaluate(df)
        return cm, acc, misclassification_rate, precision, recall, f1, roc, pr
        train_cm, train_acc, train_miss_rate, train_precision, \
            train_recall, train_f1, train_roc, train_pr = evaluation_
            metrics(train_pred_result, target_variable_name)
    test_cm, test_acc, test_miss_rate, test_precision, \
            test_recall, test_f1, test_roc, test_pr = evaluation_metrics(test_
            pred_result, target_variable_name)

    #Comparision of metrics
    print('Train accuracy - ', train_acc, ', Test accuracy - ', test_acc)
    print('Train misclassification rate - ', train_miss_rate, ', Test
    misclassification rate - ', test_miss_rate)
    print('Train precision - ', train_precision, ', Test precision - ', test_
    precision)
    print('Train recall - ', train_recall, ', Test recall - ', test_recall)
    print('Train f1 score - ', train_f1, ', Test f1 score - ', test_f1)
    print('Train ROC - ', train_roc, ', Test ROC - ', test_roc)
    print('Train PR - ', train_pr, ', Test PR - ', test_pr)

    # make confusion matrix charts
    import seaborn as sns
    import matplotlib.pyplot as plt

    def make_confusion_matrix_chart(cf_matrix_train, cf_matrix_test):

        list_values = ['0', '1']

        plt.figure(1, figsize=(10,5))
        plt.subplot(121)
        sns.heatmap(cf_matrix_train, annot=True, yticklabels=list_values,
                                    xticklabels=list_values, fmt='g')
        plt.ylabel("Actual")
        plt.xlabel("Pred")
```

365

```
    plt.ylim([0,len(list_values)])
    plt.title('Train data predictions')

    plt.subplot(122)
    sns.heatmap(cf_matrix_test, annot=True, yticklabels=list_values,
                              xticklabels=list_values, fmt='g')
    plt.ylabel("Actual")
    plt.xlabel("Pred")
    plt.ylim([0,len(list_values)])
    plt.title('Test data predictions')

    plt.tight_layout()
    return None
make_confusion_matrix_chart(train_cm, test_cm)

#Make ROC chart and PR curve
from pyspark.mllib.evaluation import BinaryClassificationMetrics

class CurveMetrics(BinaryClassificationMetrics):
    def __init__(self, *args):
        super(CurveMetrics, self).__init__(*args)

    def _to_list(self, rdd):
        points = []
        results_collect = rdd.collect()
        for row in results_collect:
            points += [(float(row._1()), float(row._2()))]
        return points

    def get_curve(self, method):
        rdd = getattr(self._java_model, method)().toJavaRDD()
        return self._to_list(rdd)

import matplotlib.pyplot as plt


def plot_roc_pr(df, target_variable_name, plot_type, legend_value, title):

    preds = df.select(target_variable_name,'probability')
```

```
preds = preds.rdd.map(lambda row: (float(row['probability'][1]),
float(row[target_variable_name])))
# Returns as a list (false positive rate, true positive rate)
points = CurveMetrics(preds).get_curve(plot_type)
plt.figure()
x_val = [x[0] for x in points]
y_val = [x[1] for x in points]
plt.title(title)

if plot_type == 'roc':
    plt.xlabel('False Positive Rate (1-Specificity)')
    plt.ylabel('True Positive Rate (Sensitivity)')
    plt.plot(x_val, y_val, label = 'AUC = %0.2f' % legend_value)
    plt.plot([0, 1], [0, 1], color='red', linestyle='--')

if plot_type == 'pr':
    plt.xlabel('Recall')
    plt.ylabel('Precision')
    plt.plot(x_val, y_val, label = 'Average Precision = %0.2f' %
    legend_value)
    plt.plot([0, 1], [0.5, 0.5], color='red', linestyle='--')

    plt.legend(loc = 'lower right')
    return None

plot_roc_pr(train_pred_result, target_variable_name, 'roc', train_roc,
'Train ROC')
plot_roc_pr(test_pred_result, target_variable_name, 'roc', test_roc, 'Test
ROC')
plot_roc_pr(train_pred_result, target_variable_name, 'pr', train_pr, 'Train
Precision-Recall curve')
plot_roc_pr(test_pred_result, target_variable_name, 'pr', test_pr, 'Test
Precision-Recall curve')
```

Save Model Objects and Create Score Code

So far, we have built a solid model and validated it using multiple metrics. It's time to save the model objects for future scoring and create score code.

- Model objects are the PySpark or Python parameters that contain information from the training process. We have to identify all the relevant model objects that would be needed to replicate the results without retraining the model. Once identified, they are saved in a specific location (HDFS or other path) and used later during scoring.

- The score code contains the essential code to run your model objects on the new dataset so as to produce the scores. Remember, the score code is used only for the scoring process. You should not have the model-training scripts in a score code.

Model Objects

In this model example, we need six objects to be saved for future use, as follows:

- `char_labels` – This object is used to apply label encoding on new data.

- `assembleModel` – This object is used to assemble vectors and make the data ready for scoring.

- `clf_model` – This object is the random forest classifier model.

- `features_list` – This object has the list of input features to be used from the new data.

- `char_vars` – character variables

- `num_vars` – numeric variables

Below code saves all the objects and pickle files required for scoring in the path provided.

```
import os
import pickle

path_to_write_output = '/home/work/Desktop/score_code_objects'
#create directoyr, skip if already exists
try:
```

```
    os.mkdir(path_to_write_output)
except:
    pass
```

```
#save pyspark objects
char_labels.write().overwrite().save(path_to_write_output + '/char_label_
model.h5')
assembleModel.write().overwrite().save(path_to_write_output + '/
assembleModel.h5')
clf_model.write().overwrite().save(path_to_write_output + '/clf_model.h5')
```

```
#save python object
list_of_vars = [features_list, char_vars, num_vars]
with open(path_to_write_output + '/file.pkl', 'wb') as handle:
    pickle.dump(list_of_vars, handle)
```

Score Code

This is a Python script file used to perform scoring operations. As mentioned before, this script file should not contain model-training scripts. At bare minimum, a score code should do the following tasks:

1) Import necessary packages

2) Read the new input file to perform scoring

3) Read the model objects saved from the training process

4) Perform necessary transformations using model objects

5) Calculate model scores using the classifier/regression object

6) Output the final scores in a specified location

First, we will create a *helper.py* script that has all the necessary code to perform scoring. The code to paste in the file is provided here:

```
#helper functions - helper.py file
from pyspark.sql import functions as F
import pickle
from pyspark.ml import PipelineModel
```

```
from pyspark.ml.classification import RandomForestClassificationModel
from pyspark.sql.functions import udf
from pyspark.sql.types import IntegerType, DoubleType

# read model objects saved from the training process
path_to_read_objects = '/home/work/Desktop/score_code_objects'

#pyspark objects
char_labels = PipelineModel.load(path_to_read_objects + '/char_label_model.
h5')
assembleModel = PipelineModel.load(path_to_read_objects + '/assembleModel.
h5')
clf_model = RandomForestClassificationModel.load(path_to_read_objects + '/
clf_model.h5')
#python objects
with open(path_to_read_objects + '/file.pkl', 'rb') as handle:
    features_list, char_vars, num_vars = pickle.load(handle)

#make necessary transformations
def rename_columns(df, char_vars):
    mapping = dict(zip([i + '_index' for i in char_vars], char_vars))
    df = df.select([F.col(c).alias(mapping.get(c, c)) for c in df.columns])
    return df

# score the new data
def score_new_df(scoredf):
    X = scoredf.select(features_list)
    X = char_labels.transform(X)
    X = X.select([c for c in X.columns if c not in char_vars])
    X = rename_columns(X, char_vars)
    final_X = assembleModel.transform(X)
    final_X.cache()
    pred = clf_model.transform(final_X)
    pred.cache()
    split_udf = udf(lambda value: value[1].item(), DoubleType())
    pred = pred.select('prediction', split_udf('probability').
    alias('probability'))
```

```
    return pred
```

After you create the file, you can score new data using the following code. You can save the scripts in a *run.py* file.

```
# import necessary packages
from pyspark.sql import SparkSession
from helper import *

# new data to score
path_to_output_scores = '.'
filename = "score_data.csv"
spark = SparkSession.builder.getOrCreate()
score_data = spark.read.csv(filename, header=True, inferSchema=True,
sep=';')

# score the data
final_scores_df = score_new_df(score_data)
#final_scores_df.show()
final_scores_df.repartition(1).write.format('csv').mode("overwrite").
options(sep='|', header='true').save(path_to_output_scores + "/predictions.
csv")
```

The final scores are available in the *final_scores_df* DataFrame, and this dataset is saved as a csv file in the final step. You need to make sure that the *run.py* and *helper. py* files exist in the same directory. That's it. We have a solid score code that is ready to be deployed in a production environment. We will go through the production implementation in the next sections.

Model Deployment Using HDFS Object and Pickle Files

This is a very simple way to deploy a model in production. When you have multiple models to score, you can manually submit each *run.py* script that you created in the previous step, or create a scheduler to run the scripts periodically. In either case, the following code will help you accomplish this task:

```
spark-submit run.py
```

As mentioned before, the *run.py* and *helper.py* files should be in the same directory for this code to work. You can configure the preceding command to specify `configurations` while submitting the job. The `configurations` are useful when you deal with large datasets, because you might have to play with executors, cores, and memory options. You can read about the `spark-submit` and `configuration` options at the following links:

```
https://spark.apache.org/docs/latest/submitting-applications.html
https://spark.apache.org/docs/latest/configuration.html
```

Model Deployment Using Docker

Let's switch gears and deploy a model using Docker. Docker-based deployment is useful in a lot of ways when compared to a traditional scoring method, like `spark-submit`.

- Portability – You can port your application to any platform and instantly make it work.

- Containers – The entire application is self-contained. All the codes and executables needed to make your application work sit within the container.

- Microservices architecture – Instead of a traditional, monolithic application, we can separate it into micro-services. This way, we can modify or scale up/down individual micro-services as and when needed.

- Faster software delivery cycles – Compatible with Continuous Integration and Continuous Development (CI/CD) processes so that application can be easily deployed to production

- Data scientists/engineers don't have to spend additional time to fix their codes/bugs when the Spark backend is altered.

- When compared to virtual machines, Docker consumes less memory, thus making it more efficient in using system resources.

Let's go ahead and implement our model using Docker. Before we start, we should have the following files in a single directory (Figure 9-1).

Name
▶ 📁 assembleModel.h5
▶ 📁 char_label_model.h5
▶ 📁 clf_model.h5
📄 Dockerfile
📄 file.pkl
📄 helper.py
📄 requirements.txt
📄 run.py

Figure 9-1. *Directory structure*

Most of the files should look familiar to you by now. We have two new files shown in the image: *Dockerfile* and *requirements.txt*. The *Dockerfile* contains the scripts to create a Docker container, and the *requirements.txt* file contains all the additional packages that you need to make your code work. Let's look at the file contents.

requirements.txt file

The contents of this file are provided here so that you can copy and paste them inside the file (Figure 9-2).

```
                requirements.txt
     pandas
     numpy
     matplotlib
     seaborn
     scikit-learn
     requests
     flask
```

Figure 9-2. *requirements.txt file*

Dockerfile

We are using the base Docker image as the PySpark Jupyter Notebook image. Inside the image, we create a working directory called /deploy/ into which we will copy and paste all our model objects and files (Figure 9-3).

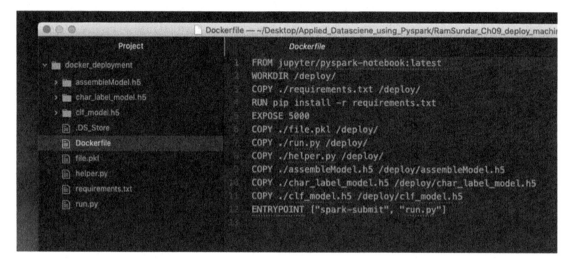

Figure 9-3. *Dockerfile*

This is done in successive steps. In addition, we install the required packages by executing the *requirements.txt* file, which is shown in line 4. Finally, we execute the model using the spark-submit command used in the previous section. One thing to notice is that the port number, 5000, is exposed in the container. We will use this port in the real-time scoring section. Just ignore this port for now. The *Dockerfile* script is provided here:

```
FROM jupyter/pyspark-notebook:latest
WORKDIR /deploy/
COPY ./requirements.txt /deploy/
RUN pip install -r requirements.txt
EXPOSE 5000
COPY ./file.pkl /deploy/
COPY ./run.py /deploy/
```

```
COPY ./helper.py /deploy/
COPY ./assembleModel.h5 /deploy/assembleModel.h5
COPY ./char_label_model.h5 /deploy/char_label_model.h5
COPY ./clf_model.h5 /deploy/clf_model.h5
ENTRYPOINT ["spark-submit", "run.py"]
```

Changes Made in helper.py and run.py Files

In the *helper.py* file, change the following line:

```
path_to_read_objects ='/deploy'
```

In the *run.py* file, change the following lines:

```
path_to_output_scores = '/localuser'
filename = path_to_output_scores + "/score_data.csv"
```

These changes are made to reflect the appropriate directory path within Docker (/deploy) and the volume that we will mount later (/localuser) during Docker execution.

Create Docker and Execute Score Code

To create the Docker container, execute the following code in the Terminal/command prompt:

```
docker build -t scoring_image .
```

Once the image is built, you can execute the image using the following command. You need to execute the command from the directory where the scoring data exists.

```
docker run -p 5000:5000 -v ${PWD}:/localuser scoring_image:latest
```

That's it. You should see the *predictions.csv* file in the local folder. Until now, we have used a batch-scoring process. Let's now move toward a real-time scoring API.

Real-Time Scoring API

A real-time scoring API is used to score your models in real-time. It enables your model to be embedded in a web browser so as to get predictions and take action based on predictions. This is useful for making online recommendations, checking credit card approval status, and so on. The application is enormous, and it becomes easier to manage such APIs using Docker. Let's use Flask to implement our model in real-time (Figure 9-4). The flask scripts are stored in the *app.py* file, the contents of which are provided next.

app.py File

```
#app.py file
from flask import Flask, request, redirect, url_for, flash, jsonify, make_
response
import numpy as np
import pickle
import json
import os, sys
# Path for spark source folder
os.environ['SPARK_HOME'] = '/usr/local/spark'
# Append pyspark  to Python Path
sys.path.append('/usr/local/spark/python')

from pyspark.sql import SparkSession
from pyspark import SparkConf, SparkContext
from helper import *

conf = SparkConf().setAppName("real_time_scoring_api")
conf.set('spark.sql.warehouse.dir', 'file:///usr/local/spark/spark-
warehouse/')
conf.set("spark.driver.allowMultipleContexts", "true")
spark = SparkSession.builder.master('local').config(conf=conf).
getOrCreate()
sc = spark.sparkContext

app = Flask(__name__)
```

```
@app.route('/api/', methods=['POST'])
def makecalc():

    json_data = request.get_json()
    #read the real time input to pyspark df
    score_data = spark.read.json(sc.parallelize(json_data))
    #score df
    final_scores_df = score_new_df(score_data)
    #convert predictions to Pandas DataFrame
    pred = final_scores_df.toPandas()
    final_pred = pred.to_dict(orient='rows')[0]
    return jsonify(final_pred)

if __name__ == '__main__':

    app.run(debug=True, host='0.0.0.0', port=5000)
```

We will also change our *Dockerfile* to copy the *app.py* file to the /deploy directory. In the last module, we ran the *run.py* file. In this module, it is taken care of by the *app.py* file. Here is the updated *Dockerfile* content:

```
FROM jupyter/pyspark-notebook:latest
WORKDIR /deploy/
COPY ./requirements.txt /deploy/
RUN pip install -r requirements.txt
EXPOSE 5000
COPY ./file.pkl /deploy/
COPY ./run.py /deploy/
COPY ./helper.py /deploy/
COPY ./assembleModel.h5 /deploy/assembleModel.h5
COPY ./char_label_model.h5 /deploy/char_label_model.h5
COPY ./clf_model.h5 /deploy/clf_model.h5
COPY ./app.py /deploy/
ENTRYPOINT ["python", "app.py"]
```

Okay, we are set to perform real-time scoring. We will code the UI shortly. Let's first test the code using the Postman application.

Recreate the Docker image using the following code:

377

```
docker build -t scoring_image .
```

You won't require volume mapping this time, since we are going to perform live scoring. You need to make sure that the port number of the local system is mapped to the Docker port number.

```
docker run -p 5000:5000 scoring_image:latest
```

```
[Sundars-MacBook-Pro:docker_deployment sundar$ docker run -p 5000:5000 scoring_image:latest
20/08/25 15:58:43 WARN NativeCodeLoader: Unable to load native-hadoop library for your platform... using builtin-java classes where applicable
Using Spark's default log4j profile: org/apache/spark/log4j-defaults.properties
Setting default log level to "WARN".
To adjust logging level use sc.setLogLevel(newLevel). For SparkR, use setLogLevel(newLevel).
 * Serving Flask app "app" (lazy loading)
 * Environment: production
   WARNING: This is a development server. Do not use it in a production deployment.
   Use a production WSGI server instead.
 * Debug mode: on
 * Running on http://0.0.0.0:5000/ (Press CTRL+C to quit)
 * Restarting with stat
20/08/25 15:59:02 WARN NativeCodeLoader: Unable to load native-hadoop library for your platform... using builtin-java classes where applicable
Using Spark's default log4j profile: org/apache/spark/log4j-defaults.properties
Setting default log level to "WARN".
To adjust logging level use sc.setLogLevel(newLevel). For SparkR, use setLogLevel(newLevel).
20/08/25 15:59:03 WARN Utils: Service 'SparkUI' could not bind on port 4040. Attempting port 4041.
 * Debugger is active!
 * Debugger PIN: 131-873-430
```

Figure 9-4. *Flask API ready*

Once you get this message, you can switch back to the Postman API and test this app. You can download Postman online and install it if you don't have it already. Initiate the application and do the necessary changes, as shown next.

Postman API

This section is intended for readers who don't have a background in the Postman API. Please skip this section if you are already familiar with it (Figure 9-5).

Figure 9-5. *Postman API*

You need to create an API instance by clicking the *New* button in the left corner and then clicking on *Request Create a basic request* to get the window shown in Figure 9-6.

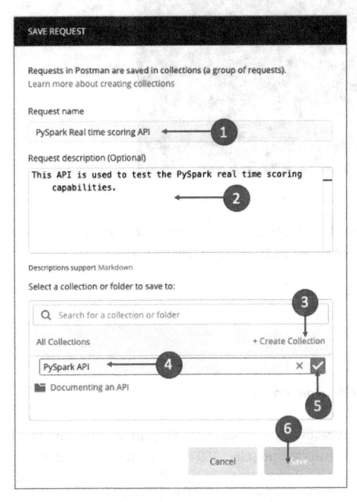

Figure 9-6. *Create a new API request.*

Now, let's make the configurations in the API shown in Figure 9-7.

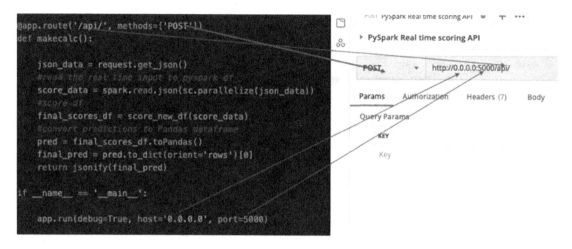

Figure 9-7. *Configurations*

Let's add a header column by navigating to the *Headers* tab. With this option, the API knows that it is looking for an JSON input (Figure 9-8).

Figure 9-8. *JSON header object*

Finally, you are ready to score. Since you are providing the input for testing, let's switch the body type to `raw` and select `JSON` as the text option (Figure 9-9).

Figure 9-9. *JSON selection*

Test Real-Time Using Postman API

Use the following JSON object for testing purposes. Note the square brackets in front. This is used to pass a list of JSON objects.

```
[{"age":58,"job":"management","marital":"married","education":"tertiary","default":"no","balance":2143,"housing":"yes","loan":"no","contact":"unknown","day":5,"month":"may","duration":261,"campaign":1,"pdays":-1,"previous":0,"poutcome":"unknown"}]
```

Copy and paste the JSON object into the body of the API and click *Send* (Figure 9-10).

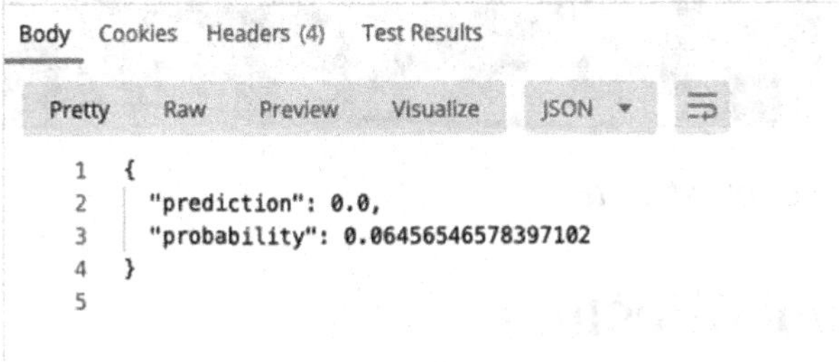

Figure 9-10.   *Real-time scores*

We can now roc proceed to build the UI component for our app.

Build UI

We will use a Python package `streamlit` to build the user interface (UI). `streamlit` is easy to integrate with our existing code. One of the greatest advantages of using `streamlit` is that you can edit the Python code and see live changes in the web app instantly. Another advantage is that it requires minimal code to build the UI. Before we get started, let's look at the directory structure shown in Figure 9-11. The parent directory is `real_time_scoring`. We have created two child directories: `pysparkapi` and `streamlitapi`. We will move all our code discussed so far to the `pysparkapi` directory. The UI code will reside in the `streamlitapi` directory.

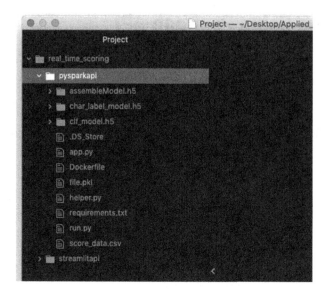

Figure 9-11. *UI directory structure*

The streamlitapi Directory

Let's navigate to the `streamlitapi` directory and create the following three files within it.

Dockerfile

```
FROM python:3.7-slim
WORKDIR /streamlit
COPY ./requirements.txt /streamlit
RUN pip install -r requirements.txt
COPY ./webapp.py /streamlit
EXPOSE 8501
CMD ["streamlit", "run", "webapp.py"]
```

requirements.txt

```
streamlit
requests
```

The following file contains the UI code.

webapp.py

```
import streamlit as st
import requests
```

```python
import datetime
import json

st.title('PySpark Real Time Scoring API')

# PySpark API endpoint
url = 'http://pysparkapi:5000'
endpoint = '/api/'

# description and instructions
st.write('''A real time scoring API for PySpark model.''')

st.header('User Input features')

def user_input_features():
    input_features = {}
    input_features["age"] = st.slider('Age', 18, 95)
    input_features["job"] = st.selectbox('Job', ['management',
    'technician', 'entrepreneur', 'blue-collar', \
     'unknown', 'retired', 'admin.', 'services', 'self-employed', \
         'unemployed', 'housemaid', 'student'])
    input_features["marital"] = st.selectbox('Marital Status', ['married',
    'single', 'divorced'])
    input_features["education"] = st.selectbox('Education Qualification',
    ['tertiary', 'secondary', 'unknown', 'primary'])
    input_features["default"] = st.selectbox('Have you defaulted before?',
    ['yes', 'no'])
    input_features["balance"]= st.slider('Current balance', -10000, 150000)
    input_features["housing"] = st.selectbox('Do you own a home?', ['yes',
    'no'])
    input_features["loan"] = st.selectbox('Do you have a loan?', ['yes',
    'no'])
    input_features["contact"] = st.selectbox('Best way to contact you',
    ['cellular', 'telephone', 'unknown'])
    date = st.date_input("Today's Date")
    input_features["day"] = date.day
    input_features["month"] = date.strftime("%b").lower()
```

```
    input_features["duration"] = st.slider('Duration', 0, 5000)
    input_features["campaign"] = st.slider('Campaign', 1, 63)
    input_features["pdays"] = st.slider('pdays', -1, 871)
    input_features["previous"] = st.slider('previous', 0, 275)
    input_features["poutcome"] = st.selectbox('poutcome', ['success',
    'failure', 'other', 'unknown'])
    return [input_features]

json_data = user_input_features()


submit = st.button('Get predictions')
if submit:
    results = requests.post(url+endpoint, json=json_data)
    results = json.loads(results.text)
    st.header('Final Result')
    prediction = results["prediction"]
    probability = results["probability"]
    st.write("Prediction: ", int(prediction))
    st.write("Probability: ", round(probability,3))
```

You can test this code before deploying in a Docker container by running the following code:

```
streamlit run webapp.py
```

As mentioned before, the interactive feature of streamlit lets you design custom layouts on the fly with minimal code. Okay, we are all set. These three files reside in the streamlitapi directory. We now have two Docker containers—one for PySpark flask and the second one for the streamlit UI. We need to create a network between these two containers so that the UI interacts with the backend PySpark API. Let's switch to the parent directory real_time_scoring.

real_time_scoring Directory

A *docker-compose* file is used to combine multiple Docker containers and create a Docker network service. Let's look at the contents of the file.

   docker-compose.yml

```
version: '3'

services:
  pysparkapi:
    build: pysparkapi/
    ports:
      - 5000:5000
    networks:
      - deploy_network
    container_name: pysparkapi

  streamlitapi:
    build: streamlitapi/
    depends_on:
      - pysparkapi
    ports:
      - 8501:8501
    networks:
      - deploy_network
    container_name: streamlitapi

networks:
  deploy_network:
    driver: bridge
```

Executing docker-compose.yml File

Let's use the following code to execute the `docker-compose.yml` file:

```
docker-compose build
```

The Docker network service is established. Let's execute the Docker service using the code shown in Figure 9-12. It will take time to spin up the service and wait for the debugger pin to appear before you switch to your browser.

```
Sundars-MacBook-Pro:real_time_scoring sundar$ docker-compose up
Creating network "real_time_scoring_deploy_network" with driver "bridge"
Creating pysparkapi ... done
Creating streamlitapi ... done
Attaching to pysparkapi, streamlitapi
streamlitapi   |
streamlitapi   |   You can now view your Streamlit app in your browser.
streamlitapi   |
streamlitapi   |   Network URL: http://172.19.0.3:8501
streamlitapi   |   External URL: http://73.81.15.250:8501
streamlitapi   |
pysparkapi     | 20/08/25 23:19:53 WARN NativeCodeLoader: Unable to load native-hadoop library for your platform... using builtin-java classes where applicable
pysparkapi     | Using Spark's default log4j profile: org/apache/spark/log4j-defaults.properties
pysparkapi     | Setting default log level to "WARN".
pysparkapi     | To adjust logging level use sc.setLogLevel(newLevel). For SparkR, use setLogLevel(newLevel).
 * Serving Flask app "app" (lazy loading)
pysparkapi     |  * Environment: production
pysparkapi     |    WARNING: This is a development server. Do not use it in a production deployment.
pysparkapi     |    Use a production WSGI server instead.
pysparkapi     |  * Debug mode: on
pysparkapi     |  * Running on http://0.0.0.0:5000/ (Press CTRL+C to quit)
pysparkapi     |  * Restarting with stat
pysparkapi     | 20/08/25 23:20:12 WARN NativeCodeLoader: Unable to load native-hadoop library for your platform... using builtin-java classes where applicable
pysparkapi     | Using Spark's default log4j profile: org/apache/spark/log4j-defaults.properties
pysparkapi     | Setting default log level to "WARN".
pysparkapi     | To adjust logging level use sc.setLogLevel(newLevel). For SparkR, use setLogLevel(newLevel).
pysparkapi     | 20/08/25 23:20:13 WARN Utils: Service 'SparkUI' could not bind on port 4040. Attempting port 4041.
 * Debugger is active!
pysparkapi     |  * Debugger PIN: 185-989-171
```

Figure 9-12. *Docker service creation*

Real-time Scoring API

To access the UI, you need to insert the following link in the browser:

`http://localhost:8501/`

The UI should look like Figure 9-13, if you used the same code as instructed in this book. We have shown only the partial UI in the figure. You can go ahead and input the values for the features and click *Get Predictions* to get results.

PySpark Real Time Scoring API

A real time scoring API for PySpark model.

User Input features

Age
18

18   95

Job

management

Marital Status

married

Education Qualification

tertiary

Have you defaulted before?

yes

Current balance
-10000

-10000   150000

poutcome

success

Get predictions

Final Result

Prediction: 1

Probability: 0.543

Figure 9-13. *Streamlit API*

When you want to kill this service, you can use the following code (Figure 9-14):

```
docker-compose down
```

```
[Sundars-MacBook-Pro:real_time_scoring sundar$ docker-compose down
Stopping streamlitapi ... done
Stopping pysparkapi   ... done
Removing streamlitapi ... done
Removing pysparkapi   ... done
Removing network real_time_scoring_deploy_network
```

Figure 9-14. *Killing Docker Service*

That's it. We have deployed our model in real-time with a nice front end. The best part of this deployment is that the UI and PySpark sit in separate Docker containers. They can be managed and customized separately and can be combined with a *docker-compose* file when needed. This makes the software development life-cycle process much easier.

Summary

- We went through the model deployment process in detail.

- We deployed a model using both batch and real-time processing.

- Finally, we created a custom UI front end using the `streamlit` API and combined with our PySpark API service using *docker-compose*.

Great job! In the next chapter, you will learn about experimentations and PySpark code optimizations. Keep learning and stay tuned.

APPENDIX

Additional Resources

In the preceding chapters, our focus was on elements specific to a particular topic in machine learning. In this appendix, however, we have stitched together some More-interesting topics that may be useful in day-to-day operations. We include a brief overview of other topics and concepts that we have not covered.

Hypothesis Testing

A hypothesis is a claim or assertion that can be tested. Based on hypothesis testing, we can either reject or leave unchallenged a particular statement. This may be more intuitive with an example. Our example claim can be that those who eat breakfast are less prone to diabetes than those who don't eat breakfast. Another example could be that the churn rate of the population who receive an offer is less than that for those who don't receive the offer. The idea is to check if these assertions are true for the population. There are two types of hypothesis, as follows:

> Null Hypothesis: This is represented by H_0. It is based on conventional wisdom or experience.

> Alternate Hypothesis: This is represented by Ha. This is often the hypothesis of interest that we are trying to validate.

If we were to formulate a hypothesis for the first of the preceding examples, we could write it as follows:

> H_0: Eating breakfast has no effect on diabetic rate.

> Ha: Eating breakfast has an effect on diabetic rate.

© Ramcharan Kakarla, Sundar Krishnan and Sridhar Alla 2021
R. Kakarla et al., *Applied Data Science Using PySpark*, https://doi.org/10.1007/978-1-4842-6500-0

The p-value is used as an alternative to rejection points to provide the smallest level of significance at which the null hypothesis would be rejected. A smaller p-value means that there is stronger evidence in favor of the alternative hypothesis. The following are types of errors:

Decision/Actual	H_0 Is True	H_0 Is False
Fail to reject null	Correct	Type II error
Reject null	Type I error	Correct

The probability of receiving a Type I error is denoted by α, and the probability of getting a Type II error is denoted by β. The power of the statistical test is given by 1-β. The cost of the Type I error far outweighs the cost of the Type II error. In most cases, we will try to reduce α as much as possible. Consider an example of flipping a coin. Effect size can be defined as the difference between heads and tails for n flips. Sample size, on the other hand, is the number of heads and tails flipped. Effect size and sample size both influence the p-value. This is the reason why we can't reduce both Type I and Type II errors at once.

The way we set up hypothesis testing is by first writing the null and alternative hypotheses. Then, we define the desired level of significance α (usually 5 percent for most tests). We run the test and make a decision based on the significance level. If the p-value is less than the significance level, we reject the null hypothesis. If the p-value is more than or equal to the significance level, we fail to reject the null hypothesis. PySpark's statistics package provides methods to run Pearson's chi-squared tests.

Chi-squared Test

The chi-square goodness of fit test is a non-parametric test that is used to find out how the observed value of a given occurrence is significantly different from the expected value. The term "goodness of fit" is used because the observed sample distribution is compared with the expected probability distribution. The test determines how well a theoretical distribution (such as Normal, Binomial, or Poisson) fits the empirical distribution.

$$\gamma 2 = \frac{(O - E)^2}{E}$$

$$\gamma 2 = Chi - square\ goodness\ of\ fit$$

O = Observed Value

E = Expected Value

The following is the sample code for verifying the p-values in a labeled dataset based on chi-squared tests.

```python
# Import Sparksession
from pyspark.sql import SparkSession
spark=SparkSession.builder.appName("HypothesisTesting").getOrCreate()

from pyspark.ml.linalg import Vectors
from pyspark.ml.stat import ChiSquareTest

data = [(0.0, Vectors.dense(0.5, 10.0)),
        (0.0, Vectors.dense(1.5, 20.0)),
        (1.0, Vectors.dense(1.5, 30.0)),
        (0.0, Vectors.dense(3.5, 30.0)),
        (0.0, Vectors.dense(3.5, 40.0)),
        (1.0, Vectors.dense(3.5, 40.0))]
df = spark.createDataFrame(data, ["label", "features"])

r = ChiSquareTest.test(df, "features", "label").head()
print("pValues: " + str(r.pValues))
print("degreesOfFreedom: " + str(r.degreesOfFreedom))
print("statistics: " + str(r.statistics))
```

The following are the p-values of the test:

```
pValues: [0.6872892787909721,0.6822703303362126]
degreesOfFreedom: [2, 3]
statistics: [0.75,1.5]
```

Kolmogorov-Smirnov Test

PySpark also has a package that supports the Kolmogorov-Smirnov test for use in basic statistics. But first, what is the KS test? It is a test of the goodness of the fit to a probability density function (PDF). It is non-parametric. It is same as the chi-square test for goodness of fit, but there is no need to divide the data into arbitrary bins. It tests whether the sampled population has the same probability density functions as another population's known or theoretical probability density functions. We'll look at a sample test.

This test can be used to check for the equality of the probability density functions of two sampled populations. This is a two-sided sample test. The hypothesis for such an occurrence can be written as follows:

$H_0 : PDF_1 = PDF_2$

$H_1 : PDF_1$ not equal to PDF_2

```
from pyspark.mllib.stat import Statistics

parallelData = spark.sparkContext.parallelize([0.1, 0.15, 0.2, 0.3, 0.25])

# run a KS test for the sample versus a standard normal distribution
testResult = Statistics.kolmogorovSmirnovTest(parallelData, "norm", 0, 1)
# summary of the test including the p-value, test statistic, and null
hypothesis
# if our p-value indicates significance, we can reject the null hypothesis
# Note that the Scala functionality of calling Statistics.
kolmogorovSmirnovTest with
# a lambda to calculate the CDF is not made available in the Python API
print(testResult)
```

Output:

```
Kolmogorov-Smirnov test summary:
degrees of freedom = 0
statistic = 0.539827837277029
pValue = 0.06821463111921133
Low presumption against null hypothesis: Sample follows theoretical
distribution.
```

Random Data Generation

Often, we would like to create data to test scenarios and algorithms. PySpark has a handy in-built function that can just do it, as shown here:

```
from pyspark.mllib.random import RandomRDDs
# Generate a random double RDD that contains 1 million i.i.d. values drawn from the
# standard normal distribution `N(0, 1)`, evenly distributed in 10 partitions.
u = RandomRDDs.normalRDD(spark.sparkContext, 1000000, 10)
# Apply a transform to get a random double RDD following `N(1, 4)`.
v = u.map(lambda x: 1.0 + 2.0 * x)
```

The preceding example generates a random double RDD, whose values follow the standard normal distribution N(0, 1), and then map it to N(1, 4).

Sampling

Sampling is one of the most important concepts in machine learning. If the sample dataset is not representative, often the experiments with biased measures carry little to no value. In this section, we will introduce types of sampling available in PySpark.

Simple Random Sampling (SRS)

In simple random sampling (SRS), every row is randomly picked and has an equal chance of being picked. The command `sample()` aids in performing SRS. We have a couple of options within SRS. For demonstration, we will use the churn modeling dataset we introduced in Chapter 8.

- SRS with replacement:

  ```
  # Read data
  file_location = "churn_modeling.csv"
  file_type = "csv"
  infer_schema = "false"
  ```

```
        first_row_is_header = "true"
        delimiter = ","

        df = spark.read.format(file_type)\
        .option("inferSchema", infer_schema)\
        .option("header", first_row_is_header)\
        .option("sep", delimiter)\
        .load(file_location)

        # Simple Random Sampling with replacement
        df=df.select('Geography','NumOfProducts','Age','Gender',
        'Tenure','Exited')
        df_srs_without_rep = df.sample(False, 0.5, 23)
        df_srs_without_rep.show(10,False)
```

- SRS without replacement:

```
        # Simple Random Sampling with replacement
        df_srs_without_rep = df.sample(True, 0.5, 23)
        df_srs_without_rep.show(10,False)
```

Stratified Sampling

In stratified sampling, every member of the population is grouped into homogeneous subgroups called strata, and a representative of each strata is chosen.

```
# Stratified Sampling

from pyspark.sql.types import IntegerType
df = df.withColumn("Exited", df["Exited"].cast(IntegerType()))
df.select('Exited').describe().show()
stratified_sampled = df.sampleBy("Exited", fractions={0: 0.5, 1: 0.5},
seed=23)
stratified_sampled.select('Exited').describe().show()
```

The following provides the raw data (left) and stratified sample distributions (right):

```
+-------+------------------+    +-------+------------------+
|summary|           Exited|    |summary|           Exited|
+-------+------------------+    +-------+------------------+
|  count|             10000|    |  count|              5022|
|   mean|            0.2037|    |   mean| 0.2078853046594982|
| stddev|0.40276858399486065|    | stddev|0.40583469642655867|
|    min|                 0|    |    min|                 0|
|    max|                 1|    |    max|                 1|
+-------+------------------+    +-------+------------------+
```

Difference Between Coalesce and Repartition

Both the `coalesce` and `repartition` functions are used to control the number of partitions used while saving the data. Here is a quick tip: If you are just archiving small- and medium-size data and are in a time crunch, use `coalesce` over `repartition`. In other cases, always try using `repartition`. Here is the reason why:

> `coalesce` uses existing partitions to minimize the amount of data that's shuffled. This results in partitions of different sizes of data.

> `Repartition` does a full shuffle and creates new partitions. This ensures equal-size partitions of data. It is also recommended you repartition your data after running filtering queries to improve the efficiency of your queries. If you do not repartition your data after filtering you may end up with many memory partitions, resulting in poor performance. You also have to keep in mind the cost of full shuffle when repartitioning data. The Directed Acyclic Graph (DAG) scheduler gives us insight into how each operation is applied on data.

Switching Between Python and PySpark

PySpark is best suited for large datasets, whereas Pandas is faster for smaller datasets. You can use Pandas data frames in concert with PySpark data frames in multiple situations to optimize the execution times of your code. We saw a classic example in Chapter 8, where we built automated pipelines. For calculating metrics, we used PySpark to do the heavy lifting, and once we had manageable data (rows in 10s to 100s with few columns), we moved them to Pandas to efficiently process and calculate metrics.

This is a great strategy when you know you will be working with rolled data cubes. You can process the large datasets in PySpark and convert the grouped PySpark data frames into Pandas data frames. Since Pandas is more mature than PySpark, doing this will open up myriad packages that can be useful in various situations.

Curious Character of Nulls

In PySpark there are situations where you are trying to fill a null in a column with data and use the `fillna` function. Your `fillna` works fine, but you still end up with nulls in a column. Yes, you guessed it right—when the datatype of the column you are trying to fill in doesn't match with the `fillna` value, you will still end up with a null value. For example, you calculate the mean for an integer column, which may end up in a float value. Now, if you are trying to fill this float value, you may still observe nulls. The workaround is to change the datatype of the column.

Common Function Conflicts

There will be instances where functions such as `max` exist in both Python and PySpark SQL functions. There will be an import conflict error if you use a wildcard import from `pyspark.sql.functions`.

Join Conditions

Joins are one of the key operations for any Spark program. In general, Spark has two cluster communication strategies. Node-to-Node involves Spark data shuffles across the clusters. Per Node performs broadcast joins. There are three main types of joins used by Spark, as follows:

- Sort Merge Join: This contains a couple of steps. First is to sort the datasets, and second is to merge the sorted data in the partition by iterating over the elements by join key. This is usually the default setting in Spark. To achieve an optimized sort merge, the join data frame should be uniformly distributed on the joining columns. If all the join keys are co-located, the join will be further optimized as there is no need for shuffle operations. Sort merge join is preferred when both sides are large.

- Broadcast Join: These joins are the most efficient ones in Spark. They are useful for joining a small dataset to huge dataset. In this join, the smaller table will be broadcasted to all worker nodes. Always make sure that you are broadcasting the smaller table.

- Shuffle Hash Join: This join is based on the concept of map reduce. This process maps through the data frames and uses the values of the join column as the output key. Spark uses a custom function to make sure the shuffle hash join is better suited for the given dataset than broadcast join. Creating a hash table is an expensive operation.

It is always good practice to verify if there are any nulls that exist in the join key. This mistake can be expensive. You may get a unrelated error such as error occurred while calling o206.ShowString or o64.cacheTable. Another reason can be duplicates. If nulls and duplicates are checked before joining tables, you can save some time by not running into any of the preceding errors.

Use sort merge join when both sides are large and a broadcast join when one side is small. These are two major strategies Spark uses for joins.

User-defined Functions (UDFs)

You may encounter issues with a UDF even after successfully importing it in the running file. Again, you can encounter a strange ShowString error. The real reason is your UDF is not available to some other worker. You can overcome this error by adding the following line to your script. It enables other workers to reference your module file.

```
sc.addPyFile(path_to_your_module.py)
```

It is also recommended that UDF usage is minimal. This because of the deserialization to reserialization operation that it triggers in the backend.

Handle the Skewness

Imbalanced datasets can cause spikes in execution duration. We can observe this in the Spark UI backend in the min and max time values. Using evenly distributed datasets can significantly improve performance.

In a SQL join operation, the join key is changed to redistribute data in an even manner so that processing for a partition does not take more time. This technique is called salting.

Add a column to each side with random integers between 0 and `spark.sql.shuffle. partitions`-1 to both sides and then add a join clause to include join on the generated column.

Using Cache

Cache is a great tool in the arsenal to speed up performance. If we use cache for all the objects it can end up in slightly slower storage. When we are using cache, it takes away some of the processing capability of your session. In the storage tab, you can observe the size in memory and size on disk distribution.

Persist/Unpersist

We know Spark uses lazy evaluation for executing commands. Persist is one of the attractive options available in the Spark store. It gives you the ability to store intermediate RDD around the cluster for faster access. For example, say I have an initial dataset that I read into Spark and to which I apply a filtering operation. Since Spark works on the concept of lazy evaluation, filtered data doesn't contain the executed data. It just maintains the order of DAG that needs to be executed for transformation. If the filtered dataset is going to be consumed by different objects to generate different results, for every single iteration the transformation occurs on filtered DataFrame (df). Imagine there are 100 operations on this filtered df. This can quickly add to the processing time and stretch the operation from taking a few minutes to taking hours for sizeable data.

Persisting data here can save a lot of computational time. When the filtered df is calculated for the first iteration, it is persisted in memory. In subsequent runs, this data frame in the memory is used for all the operations downstream. After your required downstream operations are complete, it is good practice to unpersist the stored data from memory and disk.

Shuffle Partitions

These partitions exist when shuffling data for joins or aggregations. Whenever shuffling, huge chunks of data get moved between the partitions. This shuffle can happen within the executor or across the executors. These operations can get expensive very quickly. Imagine we have a small data frame with eight initial partitions. By doing a simple group by operation, we shoot the shuffle partition count up to 200, which is the default. You have an opportunity to reset the partitions using the following configuration:

```
sparkSession.conf.set("spark.sql.shuffle.partitions",50)
```

If you are not setting these configurations, you may be underutilizing your Spark resources.

Use Optimal Formats

Spark supports many formats, such as CSV, JSON, XML, parquet, orc, and Avro. The best format for performance is parquet with snappy compression, which is the default in Spark 2.x. Parquet stores data in columnar format and is highly optimized in Spark.

Data Serialization

Spark jobs are distributed, so appropriate data serialization is important for the best performance. There are two serialization options for Spark:

- Java serialization is the default.

- Kryo serialization is a newer format and can result in faster and more compact serialization than Java. Kryo requires that you register the classes in your program, and it doesn't yet support all serializable types.

Accomplishments

- We learned the overview of hypothesis testing and different sampling techniques.

- We learned some of the common tips and tricks that can save time in day-to-day activities.

- We also had an overview of some performance fine-tuning techniques.

Fantastic job! You have completed your jour journey of this book. We hope you thoroughly enjoyed this ride. As they say, there is no end to learning. We anticipate that you will take this knowledge forward and create efficient scalable solutions for challenging questions you come across. We look forward to hearing your feedback and thoughts for improving the content of the book. Happy PySparking!!

Index

D

Printed in the United States
By Bookmasters